arkieva™

supply chain software

The Semiconductor Supply Chain:

Enterprise-Wide Planning Challenges

Ken Fordyce, PhD

Arkieva, Inc.

5460 Fairmont Drive, Wilmington DE 19808

Phone (302) 738-9215 • Fax (302) 454-7680

LIBRARY OF CONGRESS CATALOGING-IN-PUBLICATION DATA

Arkieva -

The Semiconductor Supply Chain: Enterprise-Wide Planning Challenges

ISBN-10: 0-9823148-2-5

ISBN-13: 978-0-9823148-2-1

Contents

Author's Preface

My interest in planning, scheduling, and dispatch dates back to 1972 and a position as "shift supervisor" at Friendly Ice Cream store in the brand new Nanuet Mall. Over this time period I have had the opportunity to work with a brilliant set of colleagues in a wide range of technical areas (interactive computing, data structures for fast queries, advanced planning models, algorithms for dispatching, statistical packages, analytics, parsing text, etc.) and a diverse set of application areas such as semiconductor and other manufacturing, public services, colon cancer research, not for profit foundations, customer satisfaction, public school finances; health options for seniors, and local newspapers.

Since 1977 I have focused on the successful integrated application of decision models (computational models, operations research, statistics, and artificial intelligence) and information technology (data bases, dynamic reporting, graphical user interfaces, and end user programming) to improve organizational performance by extending the borders of what Herbert Simon refers to as "bounded rationality".

Much of my work has been in planning, scheduling, and dispatch for the production of semiconductor based packaged goods – ranging from assigning lots to tools to end to end central planning. I have always seen this work as an "ongoing" challenge to reduce barriers and create new applications to support smarter decision making. The following material represents a summary of my views and thoughts based on these experiences and are divided into the following sections:

1. Enterprise demand supply networks - Core components of demand supply networks and the relationship between these components

2. The production and distribution complexity - creates challenges for planning the Production of Semiconductor Based Packaged goods

3. Factory Planning basics and challenges

4. Dispatch basics and challenges

5. Interaction between Central Planning, Factory Planning, and Dispatch

A subset of the material in a section will be repeated in another section; this is to avoid forcing the reader to go back to a prior section to review critical information.

As with any adventure, every turn of a corner or cresting a hill finds a new challenge. Over the last few years, three have emerged that are not fully covered in the following material:

- The Illusion of Capacity – Planning at an aggregated level often uses assumptions about capacity that overstates availability of resources. With increasing volatility in the market place, this disconnect has become more important but is not adequately addressed by current systems and processes.

- An improved response has made navigation and modification of the Central Plan at lot level, tool level, and flow level detail an important requirement.

- Complex process time windows – a sequence of manufacturing activities that must be completed within a certain period of time or the "lot" turns into a "pumpkin". This complexity impacts dispatch, factory planning, and central planning.

When I joined IBM in 1977 as a junior programmer with an undergraduate degree in mathematics, a few courses in operations research and statistics, one of the first people I met was Herschel Smith. Herschel was in his 60s and I was 23. Herschel had built a small linear programming model to optimize taxes for IBM World Trade. His efforts were recognized and appreciated by IBM, but he was part of an Information Systems organization. At that time, modeling did not have enough traction to be its "own person," and decision modeling applied to business problems was at most 25 years old.

Today some firms use combination of linear programming, queuing networks, and other advanced computational methods to do central planning, factory planning, and dispatch. These firms cannot think of "life without these decision models." I believe we are witnessing a growing awareness of the importance of decision models and competing on analytics [9]. What computational decision scientists offer is the potential to "Extend the Borders of Bounded Rationality."

Herbert Simon (Nobel Prize Winner in Economics) observed, "As humans, we have 'bounded rationality' and break complex systems into small manageable pieces." The challenge for organizations is to integrate information and decision technology to push boundaries out and improve performance. Nick Donofrio (IBM Senior Vice President) observed, "Access to computational capability will enable us to model things that would never have believed before." The challenge reaches beyond coding algorithms, linking to data, and turning it on. Each decision-science team must execute its role as "intelligent evolutionists" to ensure the organization adopts complex decision technology in a sustained incremental fashion. This will enable organizations from semiconductor firms to hospitals to be more responsive.

That said, there is still a long way to go and in many respects we only have the illusion of control (Kempf [51]). People are comfortable with their guesses and decision scientists often fail to deliver real value. Politely, life is much cleaner if your modeling work remains an academic exercise. The biggest obstacle may not be technology, but social order.

"There is no more delicate matter to take in hand, nor more dangerous to conduct, nor more doubtful of success, than to step up as a leader in the introduction of changes. For he who innovates will have for his enemies all those who are well off under the existing order of things, and only lukewarm support in those who might be better off under the new." (Niccolo Machiavelli)

The material in this book is based on my experience. There are many other excellent papers in the literature identifying the major challenges- a few are [24, 48, 52, 82, 83].

1. Introduction

Planning in most organizations, from health care facilities to semiconductor manufacturing, can be viewed as an ongoing sequence of loosely coupled decisions. Current and future assets are matched with current and future demand across the demand-supply network at different levels of granularity ranging from placing a lot on a tool (dispatch) to an aggregate capacity plan across a five-year horizon.

By itself, this creates a substantial challenge to management and the computational intelligence community to put in place applications that enable a firm to respond quickly and intelligently to changes in demand and/or assets.

The nature of semiconductor manufacturing adds such features as re-entrant flow, alternative bill of materials, repair actions; variability in capacity, long and short lead times, variability in demand, lead time versus utilization trade-off, to the "challenge pile".

Semiconductor manufacturing also shares key planning characteristics with process manufacturing like chemicals.

Inverted Bill of Material

In a discrete environment, many parts are used to make subassemblies which in turn are put together to make the final product; many small parts come together to make a product. For example, it takes fifty thousand or more parts to create an automobile.

Process manufacturing is the reverse. A few products or ingredients result in thousands of different products. In plastic manufacturing, ethylene may result in many different products depending on how it is processed. In semiconductors, a single wafer results in many different final parts.

Introduction

Importance of Capacity

In the process supply chains, the high investment and high technology operations are usually done early in the production process. In semiconductors, the Fab facility that does the initial processing is the single most costly step.

Because the assets that perform the processing at the start of the manufacturing process are expensive, their capacity utilization is important and often significantly affects the final product cost. In a discrete environment, the primary manufacturing issue is to make sure that the right part is available for the next production step. In process manufacturing, managing the capacity of high value assets is as important as ensuring the timely flow of product.

Non Homogenous Products

In the discrete environment, a product is either good or bad. Each part has certain specifications and if the part meets these, then it is acceptable. In the process industry, it is more complicated. When a part is manufactured, it has its own unique characteristics. In semiconductors, a unique characteristic might be the speed which the chip will tolerate. Typically, the process can be controlled so that these characteristics fall within a range. However, even in this range, the lot might be acceptable for some end-use, but not for others.

For supply chain planning, this can become complicated because the inventory of a product is not homogenous. It may be the case that there are plenty of inventories available but no inventory that will meet a particular set of requirements.

Complex Industry Dependent Issues

Facilities at the beginning of the production process tend to be capital intensive. In semiconductors, the Fab facility can run into the hundreds of billions. Because of the need to utilize these assets efficiently at the beginning of the production process, there is a desire to produce large quantities of one product before switching to another product. On the other hand, the market would like the responsiveness associated with small production runs. There is constant tension between making short runs (to speed up delivery and minimize inventory) and the need to utilize capacity effectively.

In semiconductors, products are constantly introduced. Profits and earnings are intimately tied to how quickly a product can be brought to market. Because of price erosion, a 3 to 4 month delay will result in a 30 percent reduction in total revenue for the product.

Another unique characteristic is that the manufacturing process in the Fab is highly variable. Typically, yields improve significantly over time, while prices decline.

It is also not unusual for manufacturing steps in semiconductors to be spread out over different countries. Production needs to be coordinated between these outlying facilities. This is especially challenging because the downstream processes are often sub-contracted.

Introduction

Models, Modern Computing, and Challenges

It has been approximately 50 years since the advent of modern computing and 60 years since the "general availability" of linear programming as an effective algorithm for planning (matching assets to demand over time). These two events have led to an avalanche of papers, mathematical models, and applications on planning, scheduling, and dispatch (PSD). In the last 20 years, some of these mathematical models have been incorporated into commercial systems, albeit with mixed results.

Substantial Challenges and Opportunities

All models require approximations to reality and for most of the last 50 years the core computational environment (hardware, software, and algorithms) and the business culture have imposed some "game" limiting approximations that ignore scope (complexity and variability) and scale (size).

On the "math" side we can identify such factors as:

1. lack of computational horsepower; cultural influence of the pre 1970 applied mathematics community

2. lack of access to an experimental proofing ground (the business environment); and the slow pace of diffusion

On the "business culture side", Ignizio [42] identifies two primary culprits:

1. failure to attend to the third dimension of manufacturing; the protocols (policies, practices, and procedures) employed to manage and run the supply chain and factory (the other two dimensions are: physical features and physical components)

2. lack of appreciation of the complexity, variability, and scale in the demand supply network

However, over the past 15 years the cost, in terms of lost responsiveness by failing to adequately address scope and scale in PSD processes is slowly being recognized in the business community. Many early leaders have demonstrated substantial benefits. The general conclusion is that:

- "smarter" decisions improve organizational performance

- this requires handling scope (complexity and variability) and scale (size) [52]

- which requires improved computational methods and business process

The purpose of this book is to outline the basic playing field and challenges at the global planning level, the factory planning level, the dispatch level, and co-ordination between these three levels. We hope to provide a common understanding of these challenges so that academics and companies can collaborate in developing solutions

2. Production of Semiconductor Based Package Goods

Semiconductor manufacturing is a complex process involving everything from growing silicon crystals, manufacturing silicon wafers upon which integrated circuits (ICs) are built, to the actual placement and soldering of chips to a printed circuit board ([1, 14, 49, 50]). Typically, the core manufacturing flow is wafers to devices to modules to cards (Figure 2-1).

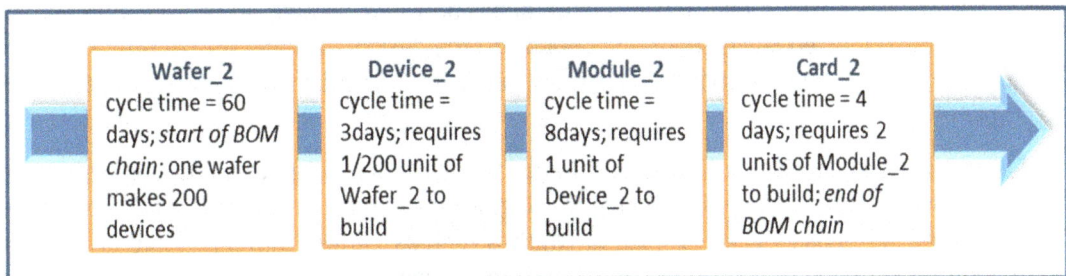

Wafer_2	Device_2	Module_2	Card_2
cycle time = 60 days; *start of BOM chain*; one wafer makes 200 devices	cycle time = 3days; requires 1/200 unit of Wafer_2 to build	cycle time = 8days; requires 1 unit of Device_2 to build	cycle time = 4 days; requires 2 units of Module_2 to build; *end of BOM chain*

Figure 2-1: Simple Flow for Production of Semiconductor Based Package Parts

2.1 Wafer Fabrication

A wafer is a thin, round slice of a semiconductor material, usually silicon. These wafers are cut from a silicon ingot. This ingot is grown from a single crystal of silicon into a long cylinder. The goal of this process is to build hundreds to thousands of identical interconnected circuits (ICs) patterns (chips) on the wafer surface according to a specific circuit design. Once the wafer is cut, circuit components are sequentially built on the wafer surface. This is referred to as the front end. These components are then connected to form the interconnected circuits (called the back end).

The front end process is broken down into five different steps:

1. Chemical clean and oxidation - prepares the wafer for photolithography

2. Photolithography - the process of "developing" the photoresist layer (photosensitive material) to expose the circuit pattern by protecting the deposited material (the mask or reticle used by the photo tools controls the exposure process in a way that corresponds to circuit structures)

3. Etching - the process of using chemicals to remove the unprotected materials in the photolithography step

4. Ion implantation - modifies the wafer's conductive properties to complete the building of circuit components and metal/insulator deposition

5. Deposition - the process of depositing materials on the wafer surface to complete the transistor

These five steps are repeated as many times as the design requires, resulting in a three-dimensional, layered structure on the two-dimensional wafer surface. The number of layers varies considerably.

The back end fabrication process is an iterative set of "metallization activities" to wire or connect the components. The individual devices are interconnected by a system of fine metal lines and vias (vertical conducting plugs between metals on different layers) using a series of photoresist-patterned metal and insulator deposition and etching steps. The number of metal-interconnected layers required depends on the design and can vary considerably. The flow in the back end is similar to the front end with "metallization" substituting for "ion implant".

Following the patterning of the last metal layer, the wafer is covered by a final dielectric layer for passivation with openings etched in this film for making electrical contacts. The wafer is then diced into individual chips, each of which is then assembled into a package or module that provides for the attachment of wire bonds to the chip.

From a PSD perspective the critical insight is the repetitive use of the same core processes across the front end and back end. The same core equipment sets are used repeatedly and this sort of flow is called "reentrant flow."

This complicates deployment decisions for partially shared operations. The following example will illustrate these issues. Figure 2-2 has a simplified set of major manufacturing activities to build one circuit layer.

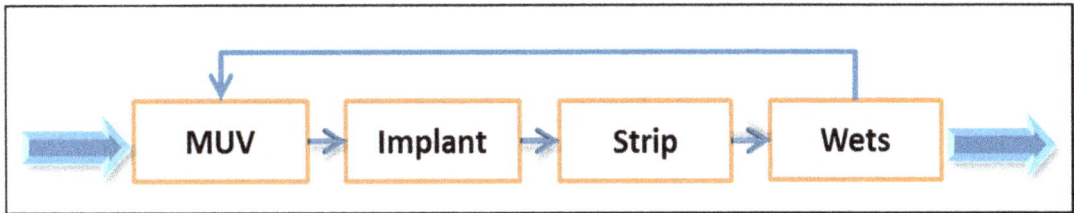

Figure 2-2 typical steps in building a circuit layer

Figure 2-3 shows the number of iterations through these same steps for Product A and Product B. In this example Product A requires three passes through the MUV toolset and Product B two passes. Assume MUV is serviced by three tools (1, 2, and 3). Typically not all tools handle all operations (passes).

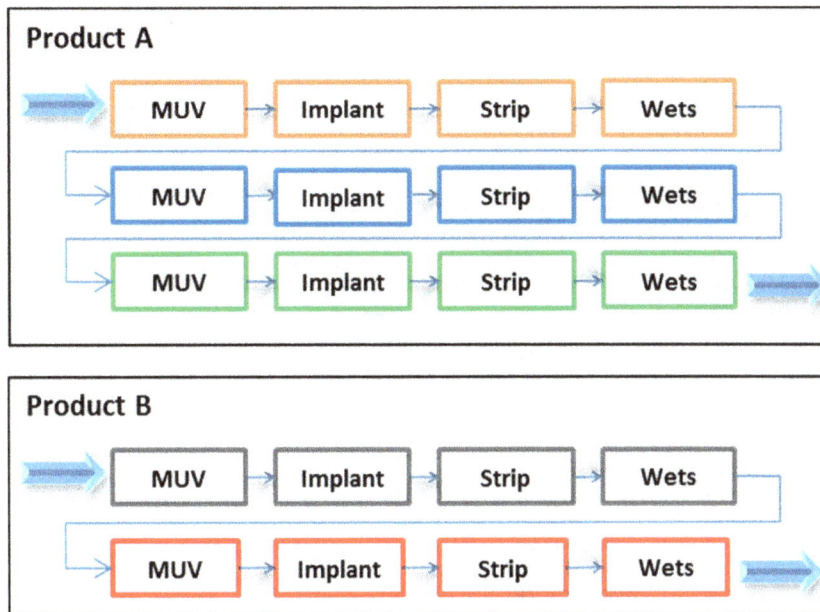

Figure 2-3: Example of reentrant flow for typical circuit construction

Figure 2-4 demonstrates the concept of partially shared tools sets for related, but different, manufacturing operations. In this example there are three tools (1, 2, and 3) to handle the MUV step. But not all tools can handle all steps. This is the deployment decision.

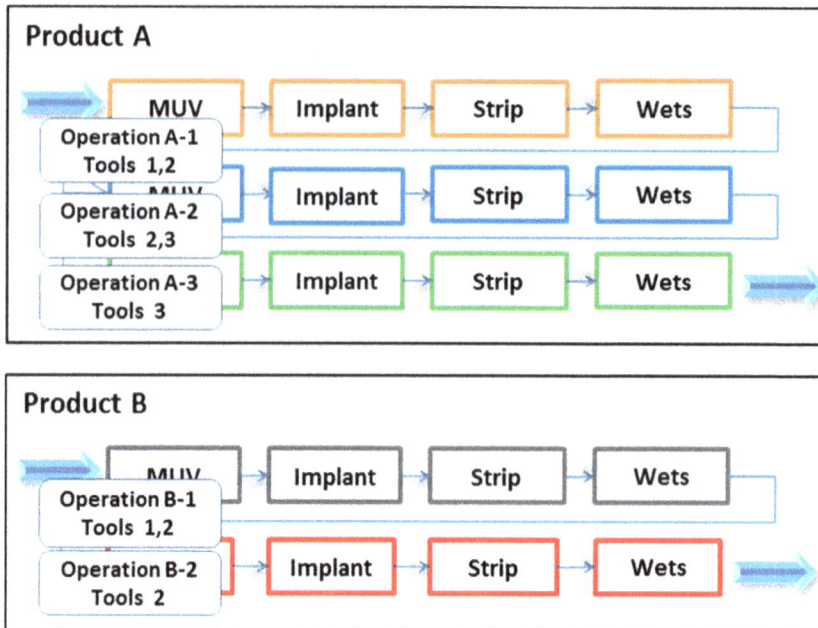

Figure 2-4 Reentrant flow with Partial Shared Operations (PSO)

In table form, the deployment decision would be represented as follows:

	Tool 1	Tool 2	Tool 3	Number of tools Covering Operation
Operation A-1	1	1	0	2
Operation A-2	0	1	1	2
Operation A-3	0	0	1	1
Operation B-1	1	1	0	2
Operation B-2	0	1	0	1
Number of Operations that the tool covers	2	4	2	

Table 2.1 Deployment Information for Circuit Layers

The decision about which tools can handle which operations (deployment decision) is determining which cells in table 2.1 get a 1 or a 0.

For complex designs, it may take three to six hundred individual steps to complete the entire IC fabrication process. Cycle times range from 30 to 130 days. The yield (percentage of quality wafers) can show significant variation.

There is also a wide variation in the demand patterns for wafers. A wafer that is a component of a video game has large demands over a 6-to-12-month period. In contrast, wafers that support specialty products have sporadic demands over a longer time horizon. In contrast, wafers used in sensing and testing devices generally have stable demands.

Wafers used in computers or control units have moderate demands over a reasonable period of time and then fall into an end of life phase that must cover warranty issues.

There is also no uniformly accepted way to specify demand. Supply chain professionals and the business often states demand as the finished wafers required, while factory planners refer to demand as the number of planned wafer starts.

For all of the complexity and uncertainty in wafer fabrication, all enterprise planning engines have a simple representation of this manufacturing process: a few key capacity points and a limited number of purely serial decision points.

For additional details of the dispatch and scheduling complexity in wafer fabrication, see [43, 49]

2.2 Wafer Testing and Device Generation (Binning)

Once the circuits have been built on the wafers, they are tested to determine the resultant yield of operational circuits (good or bad, speed, power consumption, etc.) and tagged for reference. The wafers are then diced (cut) into individual units and sorted or binned based on the prior testing, creating what is commonly referred to as the "die bank."

This process is generally referred to as "binning" (illustrated in Figure 2-5). Observe it is the device (not the wafer) that is packaged and placed into video games, cell phones, laptops, etc.

A single wafer generates anywhere from 30 to 900 devices, and sometimes, the exact type of device is not determined until it is tested. The testing not only determines whether the device is usable or not, but also its final part identity. Differences between devices from the same wafer typically occur as result of speed and power consumption. Wafer testing has a short cycle time (3 to 10 days) and, except for rework, involves a purely sequential process through test operations.

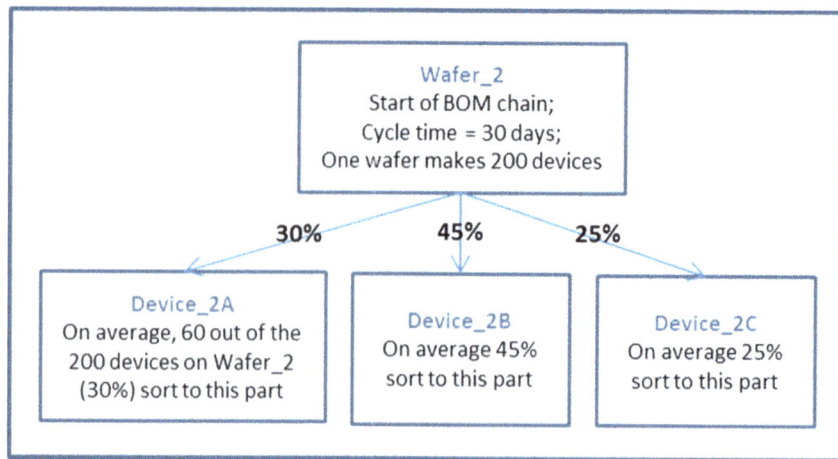

Figure 2-5: Simple Binning or Sorting Situation from Wafer to Device

2.3 Modules

The device step is followed by a sequence of assembly steps to mount devices onto a substrate (called wire bonding) and package to make a module. Packaging protects the fragile device inside and makes it suitable for incorporation in other electronic products. These modules are further tested to determine electromagnetic and thermal characteristics.

As with devices, in some cases the final identity of the part may not be determined until the test activities are done. Again, the cycle times are short compared to fabrication times.

For some modules, a significant amount of complexity is introduced when the production process requires a sequence of packaging and testing steps and/or the same module can be made from different processes that may or may not consume the same component part (a.k.a. alternate build paths). Figure 2-6 provides an example of this type of complexity.

2.4 Cards

The modules are eventually combined onto printed circuit boards to make cards. The cards are tested and those that pass are used in the assembly of a wide range of electronic products such as PCs, printers, and CD players. The cards are often produced at multiple vendor locations.

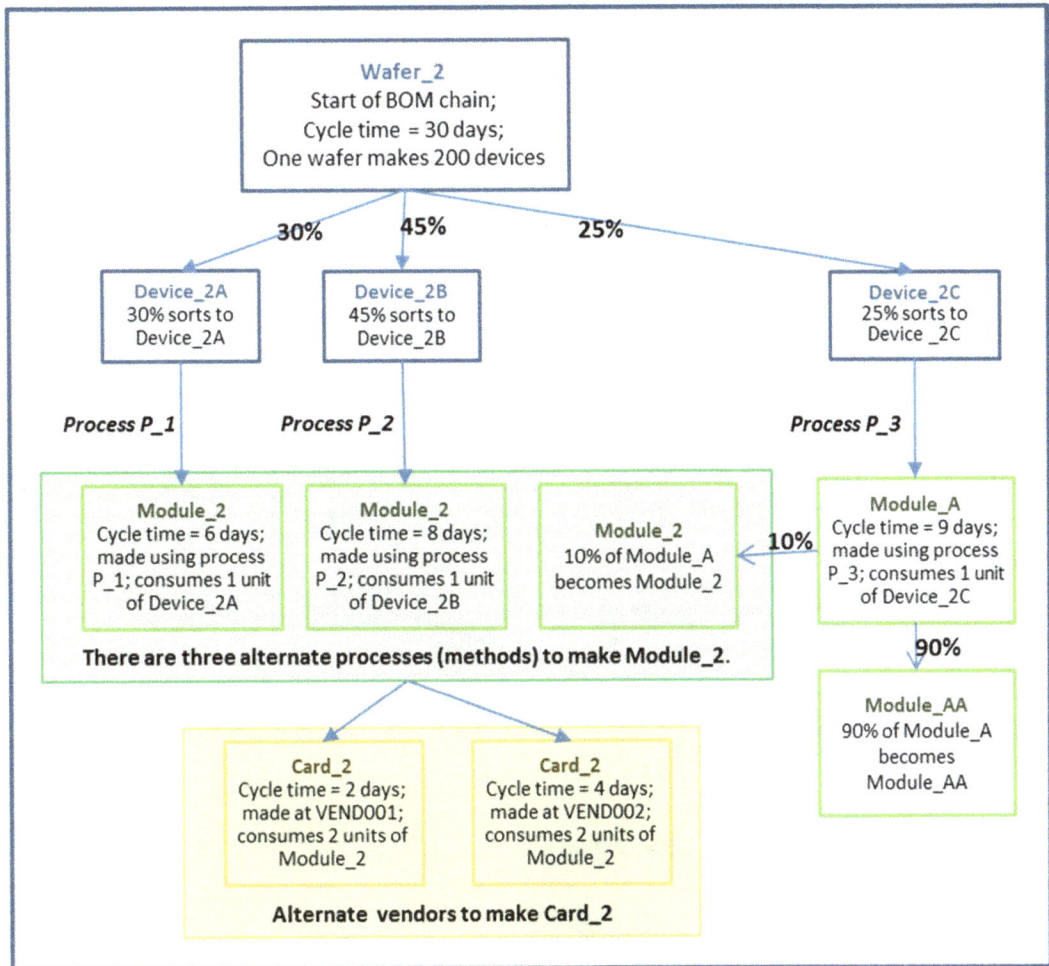

Figure 2-6: Alternative Processes to Make Module_2, Module Binning, and Card_2

2.5 Total Journey

To create manageable enterprise wide supply chain models, we need to simplify the process by abstracting the complex routing of lots through the production process. Although the products have continuous properties, we segregate these into a discrete set of part numbers (PNs). We use a bill of material (BOM) structure that specifies the component(s) of each PN so that we can generate a graphical representation of the components needed for finished products. The alternative paths to produce the same PN and the binning or sorting activities that determine the final status of a PN are illustrated in (Figure 2-7).

The total journey from sand to module or card is inherently complex. The first stage (fabrication of wafers and then testing and sorting into devices) has a long lead or cycle time where the capital expense for equipment ranges from hundreds of millions to billions of dollars.

There is no assembly at this stage; the number of PNs goes from a few to many. In stage two (modules) and three (cards), the cycle times are short and the process is assembly and test. Typically, more than one component is needed in the assembly process and there are multiple ways to produce the same part.

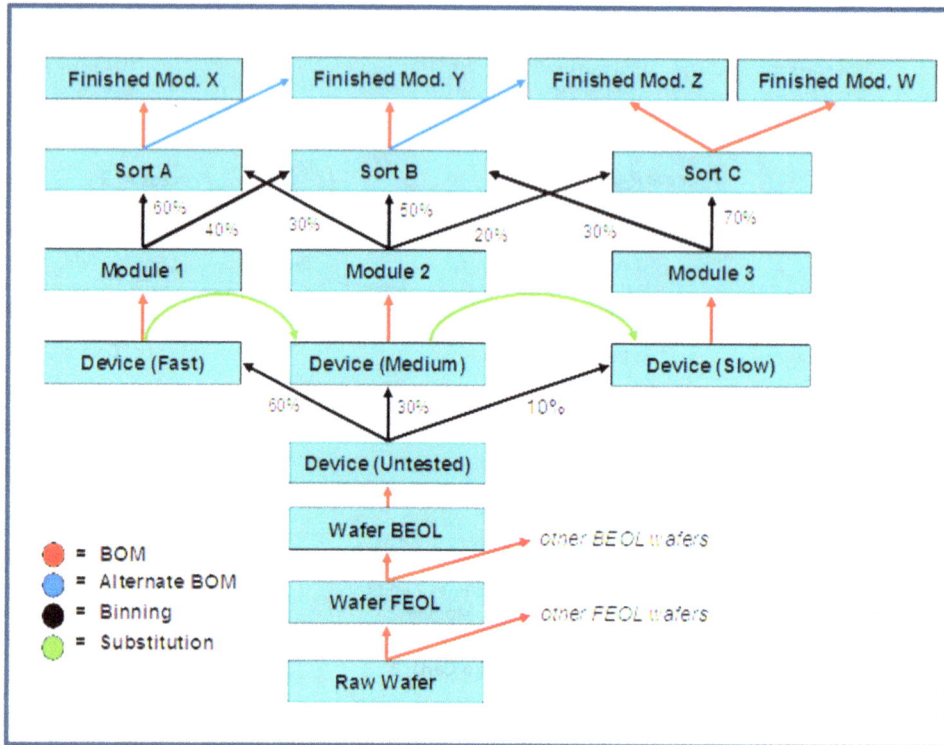

Figure 2-7: Graphical Representation of a Bill of Material Structure

The number of finished goods explodes compared to the original number of wafer parts. A manufacturing facility typically specializes in only part of the process. Example: the wafer is produced at location A; it is tested and diced at location B; the module is assembled at location C1, C2, or C3; it is tested at location D1 and then sent to location E1 or E2 to be assembled onto a card or sent directly to a customer. Within this network, different PNs that support a wide range of customers are constantly competing for the same tool or machine capacity.

The complexity extends to variability across time as well. Most of the planning parameters and business preferences are date or time effective. This includes cycle times, yields, capacity required, capacity available, binning percentages, manufacturing processes, and so on. For example, in Figure 2-6 the cycle time to produce Module_2 might reduce from 8 to 5 days in three weeks, and the entire process might be eliminated in 30 weeks.

To add to the complexity, customer or exit demand may occur at any level (card, module, device, wafer tested, and wafer untested), or be stated as required starts or allocation of capacity.

For example, a customer might place its primary demand as wafers and an additional "value add" demand for the firm to take some of the chips from the finished wafers and turn them into modules. In this case, the demand for finished wafers has a higher priority than the value add demand. If the firm doesn't fully satisfy the value add demand, the customer could have that work done by another vendor.

3. Levels and Components of the Playing Field

Organizations, from health care facilities to manufacturing giants to small restaurants, can be viewed as an ongoing sequence of loosely coupled activities where current and future assets are matched with current and future demand across the demand-supply network.

These planning, scheduling, and dispatch decisions across a firm's demand-supply network are best viewed as a series of information flows and decision points organized in a decision hierarchy or tiers. The row dimension is the decision tier and the column is the responsible unit. Observe that the decisions in each tier limit the options in the tiers below it.

The first decision tier, strategic planning, is typically driven by the lead time required for business planning, resource acquisition, and new product introduction. Here, decision makers are concerned with a set of problems that are three months to seven years into the future. Issues considered include, but are not limited to, what markets the firm will be in, general availability of equipment and skills, major process changes, risk assessment of changes in demand for existing products, required or expected incremental improvements in the production or delivery process, and the lead times for adding additional equipment and skills.

The second tier, tactical planning, deals with problems the company faces in the next week to six months. Estimates are made of yields and cycle times (lead times), the likely profile of demand, productivity, reliability of equipment, etc. Decisions are made about scheduling releases into the manufacturing line or staffing levels. Delivery dates are estimated for orders or response times for various classes of patients is estimated. Deployment of equipment and staffing is adjusted, the order/release plan is generated or regenerated, and (customer-requested) reschedules are negotiated.

	Enterprise wide, Global View – Central Planning	Enterprise sub-units (manufacturing, distribution, retail) - Factory Planning
Tier 1 Strategic	Enterprise wide Central Planning once or twice a year for a 2 to 5 year horizon at an aggregated level with forecasted demand and focused on business scenarios. Net result is a strategic direction is established and financial commitments are made.	Capacity analysis typically at the tool/family level and overall manpower to support forecasted demand; creation of production flow and capacity information for central plan; determining new production processes to introduce , and estimating the learning curve.

	Enterprise wide, Global View – Central Planning	Enterprise sub-units (manufacturing, distribution, retail) - Factory Planning
Tier 2 Tactical	Enterprise wide Central Planning weekly/biweekly/monthly • Create demand statement (current orders, forecasts) • Capture capacity, WIP, BOM, business policy • Central planning engine to match assets with demand • Estimate supply line linked to demand, monitor early warning signals, chase potential concerns	Capacity (tools and manpower) analysis to gauge the impact of changing product mix, identity challenges, review and modify deployment decisions, and manufacturing engineering requirements, and create capacity constraint information for central planning and WIP status. Monitor tool level performance and take appropriate actions. Establish rules and metrics to set global lot importance – example, how many priority classes, algorithm to set lot importance within a class, limits on number of expedites.

The third tier, operational scheduling, deals with the daily execution and achievement of a weekly plan. Shipments are made, patients receive treatments, customers are waited on, serviceability levels are measured, and recovery actions are taken. Optimal capacity consumption and product output are computed.

Tier "3.5" straddles operational and real time response. For example, a monitoring system might observe a lot has entered a "process time window" and its "urgency" to be assigned has "increased". A process time window is a sequence of activities that must be accomplished within a certain time limit or the lot might need to be scrapped due to some type of contamination. A non-factory example would be the "triage" system that occurs regularly in an emergency room where a patient is placed in one of four or five categories based on urgency. Although this decision does not directly assign the patient to a health care provider, it has a strong influence over the type and urgency of the assignment.

	Enterprise wide, Global View – Central Planning	Enterprise sub-units (manufacturing, distribution, retail) - Factory Planning
Tier 3 Operational (daily)	Enterprise wide Central Planning has reduced focus. What if analysis: • What-if commit on large orders • What-if for major asset changes • Status of key WIP and actions to take • Cross-factory signals	Provide information to central plan and daily factory adjustments. • Establish target outs and due dates on lots • Maintenance priorities • Short term changes in deployment • Review key lot status and change priority (up or down) based on progress (either manually or dynamically) • One time changes in lot importance guidance • Establish manufacturing lot versus development lot preference • Revised projected outs for enterprise planning

	Enterprise wide, Global View – Central Planning	Enterprise sub-units (manufacturing, distribution, retail) - Factory Planning
Tier 3.5 Sub-daily guidance	Change in priorities; updated supply projections based on updated WIP or capacity status; change in customer reserved supply.	As needed updates to guidance to support response decisions. • Regular updates to lot status based on its progress, entering a time process window, status of short term manufacturing targets, WIP position, and tool status. • Regular updates to tool status based on manufacturing engineering requirements, tool events, etc.

The fourth tier, real-time response system, addresses the problems of the next hour to a few weeks by responding to conditions as they emerge. Within the demand-supply network, real time response is often found in two areas: manufacturing dispatch (assign lots to tools) and order commitment (available to promise, or ATP).

Within manufacturing, the decisions made across the tiers are typically handled by groups with one of two responsibilities: maintaining an enterprise wide global view of the demand-supply network and ensuring that subunits (such as manufacturing location, vendor, warehouse, etc.) are operating efficiently.

Ideally all planning would be centralized; in practice, complexity precludes this. Capacity planning is a good example of this. At the enterprise level, capacity is modeled at some level of aggregation, typically viewing a key tool set as a single capacity point or simply based on starts. At the factory level, each tool, or potentially each chamber in a tool, is modeled.

	Enterprise wide, Global View – Central Planning	Enterprise sub-units (manufacturing, distribution, retail) - Factory Planning
Tier 4 Response	Available to promise or automated order commit process. Cross factory signals	Dispatch schedule and tool response • Assign sequence of lots to a tool • Change the status of a lot (for example on or off hold) • Monitor signals from tools and respond as needed

Level and Components of the Playing Field

Demand Supply (DS) Network Planning, Scheduling, and Dispatch (PSD) Activity Areas and Decision Tiers		
Decision Tiers	**Enterprise wide, Global View – Central Planning**	**Enterprise sub-units (manufacturing, distribution, retail) - Factory Planning**
Tier 1 Strategic	Enterprise wide Central Planning once or twice a year for a 2 to 5 year horizon at an aggregated level with forecasted demand and focused on business scenarios. Net result is that a strategic direction is established and financial commitments are made.	Capacity analysis typically at the tool/family level and overall manpower to support forecasted demand; creation of production flow and capacity information for central plan; determining new production processes to introduce , and estimating the learning curve.
Tier 2 Tactical	Enterprise wide Central Planning weekly/biweekly/monthly • Create demand statement (current orders, forecasts) • Capture capacity, WIP, BOM, business policy. • Central planning engine to match assets with demand • Estimate supply line linked to demand, monitor early warning signals, chase potential concerns	Capacity (tools and manpower) analysis to gauge the impact of changing product mix, identity challenges, review and modify deployment decisions, and manufacturing engineering requirements, and create capacity constraint information for central planning and WIP status. Monitor tool level performance and take appropriate actions. Establish rules and metrics to set global lot importance – example, how many priority classes, algorithm to set lot importance within a class, limits on number of expedites.
Tier 3 Operational (daily)	Enterprise wide Central Planning has reduced focus. What if analysis: • What-if commit on large orders • What-if for major asset changes • Status of key WIP and actions to take • Cross-factory signals	Provide information to central plan and daily factory adjustments. • Establish target outs and due dates on lots. • Maintenance priorities • Short term changes in deployment • Review key lot status and change priority (up or down) based on progress (either manually or dynamically) • One time changes in lot importance guidance. • Establish manufacturing lot versus development lot preference. • Revised projected outs for enterprise planning
Tier 3.5 Sub-daily guidance	Change in priorities; updated supply projections based on updated WIP or capacity status; change in customer reserved supply.	As needed updates to guidance to support response decisions. • Regular updates to lot status based on its progress, entering a time process window, status of short term manufacturing targets, WIP position, and tool status. • Regular updates to tool status based on manufacturing engineering requirements, tool events, etc.
Tier 4 Response	Available to promise or automated order commit process. Cross factory signals	Dispatch schedule and tool response • Assign sequence of lots to a tool • Change the status of a lot (for example on or off hold) • Monitor signals from tools and respond as needed

20

3.1 Enterprise Wide Central Planning

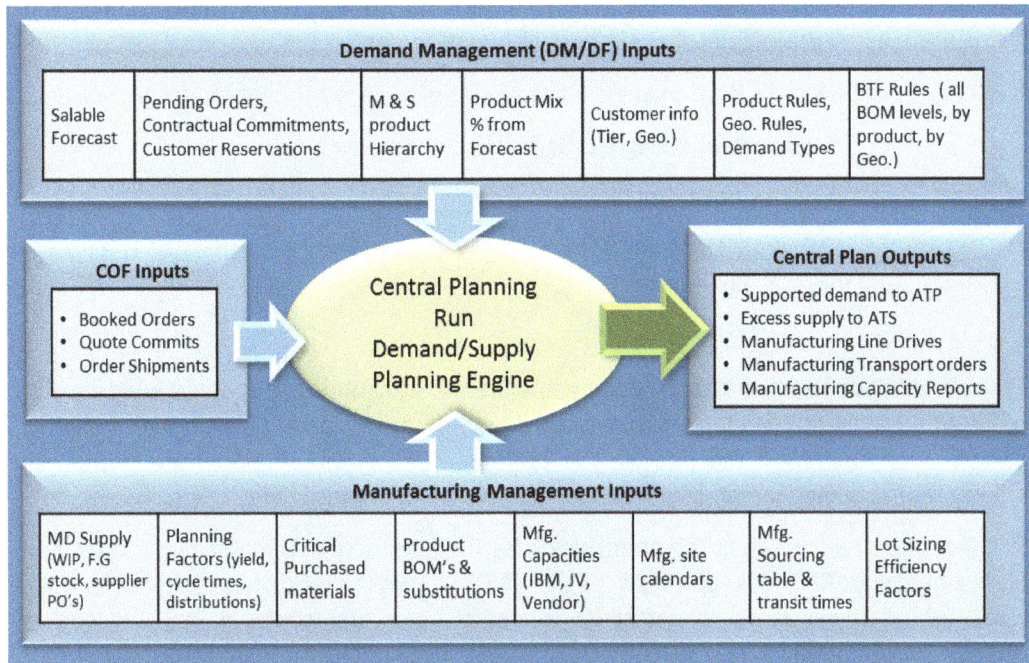

Demand Management (DM/DF) Inputs						
Salable Forecast	Pending Orders, Contractual Commitments, Customer Reservations	M & S product Hierarchy	Product Mix % from Forecast	Customer info (Tier, Geo.)	Product Rules, Geo. Rules, Demand Types	BTF Rules (all BOM levels, by product, by Geo.)

COF Inputs
- Booked Orders
- Quote Commits
- Order Shipments

Central Planning Run Demand/Supply Planning Engine

Central Plan Outputs
- Supported demand to ATP
- Excess supply to ATS
- Manufacturing Line Drives
- Manufacturing Transport orders
- Manufacturing Capacity Reports

Manufacturing Management Inputs							
MD Supply (WIP, F.G stock, supplier PO's)	Planning Factors (yield, cycle times, distributions)	Critical Purchased materials	Product BOM's & substitutions	Mfg. Capacities (IBM, JV, Vendor)	Mfg. site calendars	Mfg. Sourcing table & transit times	Lot Sizing Efficiency Factors

Figure 3-1: Typical Data Inputs and Outputs for Enterprise-wide Central Planning Engine

This activity involves the coordination of supply and demand (current and future) [17, 75] across each unit in the enterprise to determine which customer commitments can be made and what opportunities to chase. It consists of these core activities and illustrated in Figure 3-1

1. Create a demand statement
2. Capture the flow of materials in the demand supply network
3. Gather and collect key information from the factory
    a. Project the completion of WIP to a decision point (often completion of the part).
    b. a statement of required and available capacity
    c. a statement of lead time or cycle time to complete a new start
4. Create a model that captures key enterprise relationships (Central Planning Engine – CPE)
5. Create an enterprise wide central plan by matching current and future assets with current and future demand using the CPE to create a future projected state of the enterprise and the ability to soft peg the current position of the

enterprise to the projected future position. Information from the model includes:

a. a projected supply linked with exit demand

b. identification of at-risk orders either by commit date or request date

c. synchronization signals across the enterprise

d. capacity utilization levels

e. ability to trace each production & distribution activity that supports meeting a demand

Typically this is an iterative process done with different assumptions about available capacity, different business policies for protective stock, and different commit dates and/or demand priorities for orders. At the end of this process, the results are communicated to the organization including the projected supply to available to promise (ATP).

Step 4 is often referred to as the enterprise wide best-can-do (BCD) matching, or the central planning engine (CPE). The core task of the CPE is to deploy modeling methods to match assets with demand across an enterprise to create a projected supply linked with demand and synchronization signals. Assets include, but are not limited to, manufacturing starts (or releases), work in progress (WIP), inventory, purchases, and capacity (manufacturing equipment and manpower). Demands include, but are not limited to, firm orders, forecasted orders, and inventory buffer. The CPE has four core components:

- Represent the (potential) material flows in production, business policies, constraints, and demand priorities, current locations of assets, and relate all this information to exit demand

- Capture asset quantities and parameters (cycle times, yields, binning percentages, etc.)

- Search and generate a supply chain plan, relate the outcome to demand, and modify the plan to improve the match

- Display and explain the results

This flow is summarized in Figure 3-2.

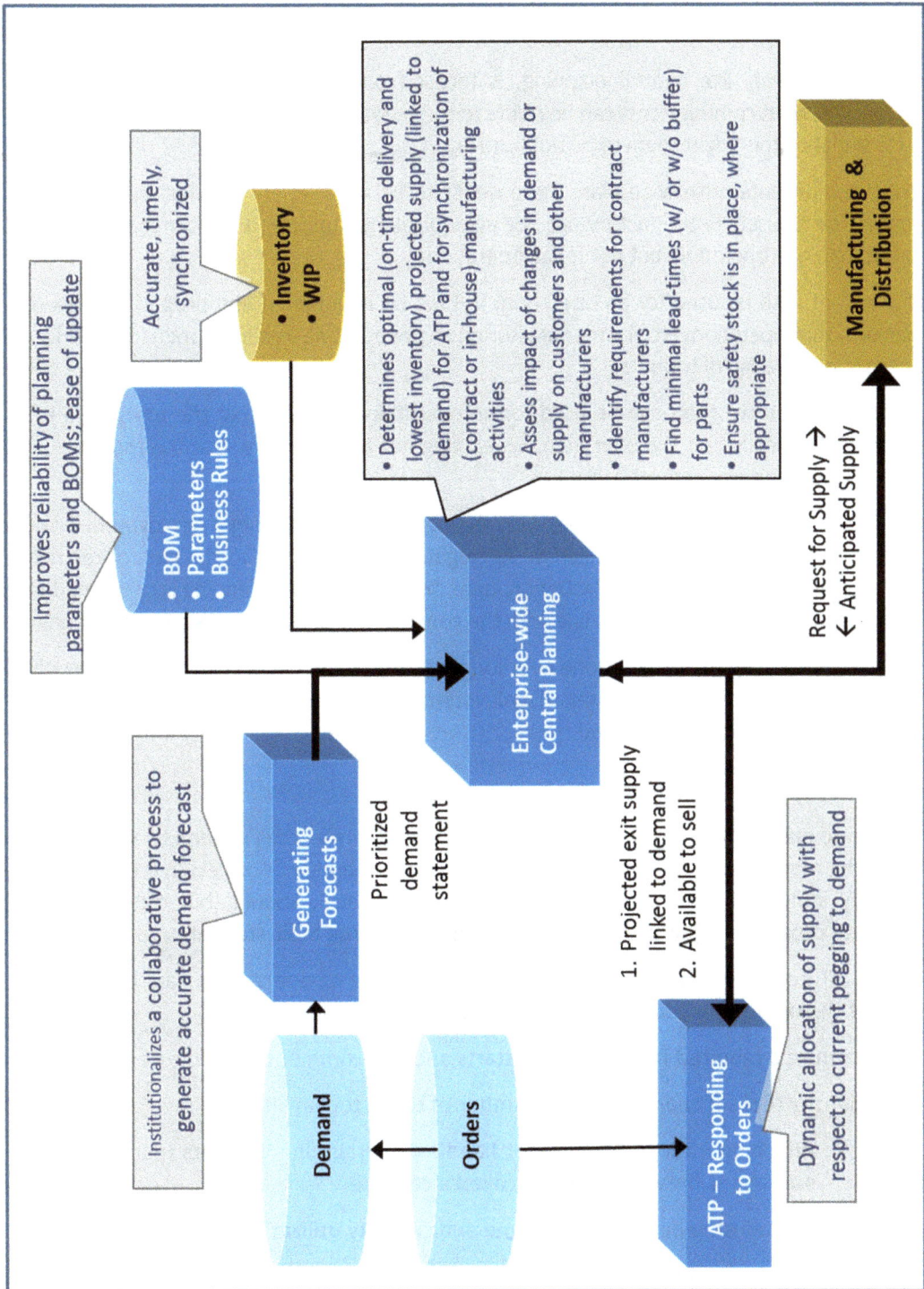

Figure 3-2: Enterprise-wide Central Planning Engine Flow

3.2 Basics of Factory Planning

Factory planning, like central planning, is focused on matching assets with demand to determine what commitments can be made to the business and what opportunities to chase. At first glance this might seem to be duplicative.

To create an enterprise wide plan, there needs to be a statement of what the factory can make. The toolsets in the factory are not all equivalent. Typically a toolset can be used for more than one operation, but not for all operations.

If a toolset is to be used for an operation, it must be maintained and properly qualified to perform that operation. From a manufacturing engineering viewpoint, workload is reduced if there is less overlap of toolsets.

For example, if two tools can be used for an operation, it means that the tools have to be individually qualified on the operation. Scheduling has to make sure that the tools are used for the operation on a regular basis.

There are typically three major types of planning: aggregate planning; deployment or near term tool planning, and WIP Projection. Aggregate planning refers to the step of restricting toolsets to certain operations only. Deployment or near term tool planning is concerned with assigning workload to the tools in line with the constraints established in aggregate planning.

Typically, aggregate planning is done over longer time horizon because of the effort involved in qualifying and preparing toolsets. Deployment is dynamic and is done as often as the central planning engine is executed.

3.2.1 Aggregate Planning

Aggregate planning is typically focused on calculating the resources needed by the factory to meet the requirements of central planning. It identifies "broken" toolsets and estimates the capacity inputs required by the central planning model. It should (but often does not) integrate capacity analysis with the cycle time estimates. The basic steps are:

1. Capture representative product routes: sequence of operations, raw process time at the operation, and tool set consumed for each key operation

2. Capture a required factory load in starts and lead-time or cycle time commit

3. Identify tool set characteristics: number of tools; tool availability

   3.1. Which operations the tool set handles, overlapping tool sets (partially shared operations between toolsets called a cascade

   3.2. Relationship between cycle time and capacity utilization

   3.3. Reentrant flow (when a lot visits the same tool set multiple times)

4. Allocate tool sets to parts or families either as user input or based on history

5. Execute a model to determine how this load impacts toolsets

    5.1. Required utilization levels for each tool set to meet demand stated as starts including identifying broken tools (tools without sufficient capacity to meet demand)

    5.2. Provide some details on which parts are consuming how much of a tool set

    5.3. Capacity loss points (planned maintenance, high raw process time, etc.)

6. Provide information for central planning

    6.1. High level statement of capacity based on starts or a few key tool sets fixed for the specified cycle time (serves to limit the requirements placed on the factory by central planning)

    6.2. More starts or lower cycle time

7. Identify the additional capacity needed for the factory to handle

Unlike Central Planning, aggregate factory planning deals with:

- The actual route (sequence of manufacturing actions to produce a finished part)

- The trade-off between cycle time and tool utilization / capacity available.

- Partially Shared Operations (PSO) or Cascade which refers to different toolsets that overlap in the manufacturing operations they service, but the actions are not identical.

- Deployment Variability

  - Product mix transitions

  - Reentrant flow

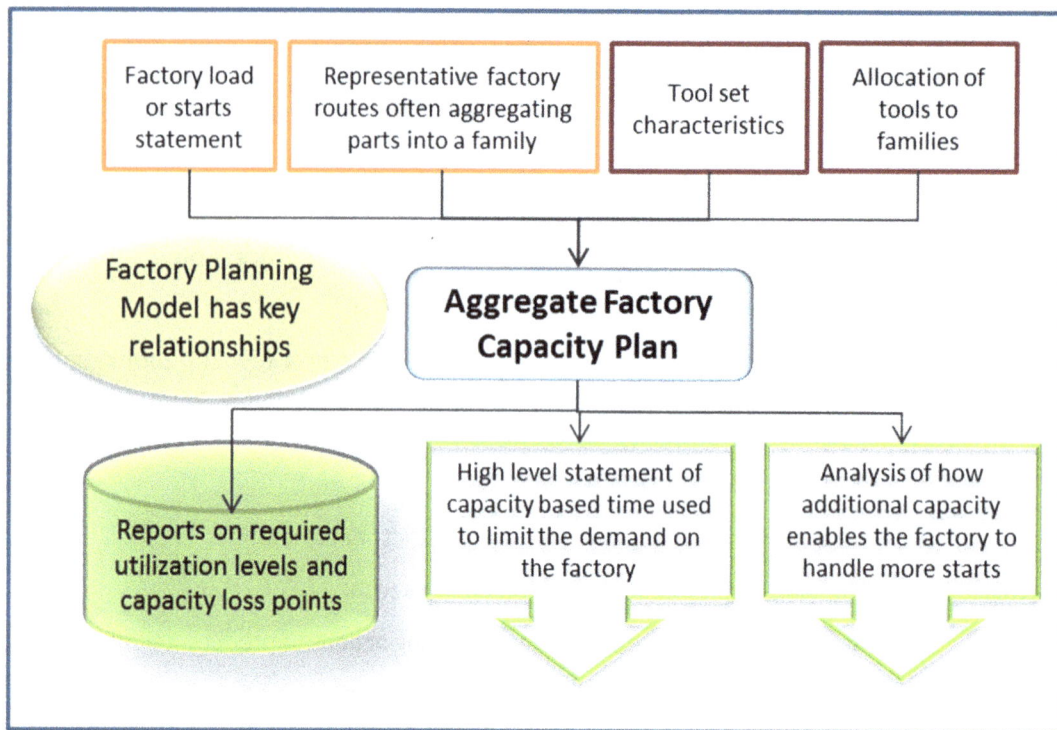

Figure 3-4 Basics of Aggregate Factory Planning

3.2.2 Deployment or Near Term Tool Planning

In Table 3.1 we have three tools (1, 2, and 3) to service seven operations (001 – 007) creating a 7 by 3 grid. A 1 indicates that this tool can service this operation and 0 it cannot. This is called PSO or Cascade group since there are no tool/operation subsets that are independent – at some level each of these operations and tools are connected. From a tool perspective:

- Tool 1 and Tool 2 "share" operations 001 and 002

- Tool 1 and 3 do not share any operations

- Tool 2 and Tool 3 share operations 003 and 005

- Therefore the connection between Tool 1 and Tool 2 is a second order link.

	Tool 1	Tool 2	Tool 3	Number of tools Covering Operation
Oper001	1	1	0	2
Oper002	1	1	0	2
Oper003	0	1	1	2
Oper004	0	0	1	1
Oper005	0	1	1	2
Oper006	1	0	0	1
Oper007	1	0	0	1
Number of Operations that the tool covers	4	3	2	

Table 3.1: Deployment Information for PSO Group

Table 3.2 has the same tools and operations as table 3.1, but with a different deployment allocation. In this situation there is NO overlap in operations between Tool 2 and Tool 3, therefore there are two PSO groups. Tool 1 and Tool 2 (and the operations they service – 1, 2, 5, 6, and 7) and Tool 3 (operations 3 and 4).

	Tool 1	Tool 2	Tool 3	Number of tools Covering Operation
Oper001	1	1	0	2
Oper002	1	1	0	2
Oper003	0	0	1	1
Oper004	0	0	1	1
Oper005	0	1	0	1
Oper006	1	0	0	1
Oper007	1	0	0	1
Number of Operations that the tool covers	4	3	2	

Table 3.2: Deployment Information for 2 PSO Groups

An additional wrinkle to this feature is the raw process times for the same operation may differ between tools (Table 3.3).

The near term deployment decision is to determine which cells in table 3.1 get a 1 and which get a 0. Typically a tool can service many operations, but a factory will limit the number of operations it is "allowed" to service in order to:

- Improve yield and speed by focusing tools on the operations they are best suited to handle

- Ensure a tool remains qualified to handle an operation by regularly servicing this operation

- Account for tool specific restrictions which may limit a tool from serving some operations based on the operations it can currently service

- Reduce work load on manufacturing engineering to maintain tool qualification

- Simplify dispatching by reducing options

	Tool 1	Tool 2	Tool 3
Oper001	4	20	5
Oper002	15	20	6
Oper003	10	15	8
Oper004	10	9	20
Oper005	5	5	5
Oper006	8	10	10
Oper007	10	10	10

Table 3.3: RPT Information

3.2.3 WIP Projection

Part of the input to the central planning engine is when the current WIP will be available for use. All central planning engines require the factory to project when the current WIP will be completed or reach a staging point. The typical process involves capturing the remaining steps (route) for the lot and estimating how long it will take to finish these steps.

3.3 Factory Floor Dispatch

Dispatch refers to assigning a lot to a tool and requires balancing effective tool utilization with stable delivery (either to the commit date on the lot or to the number of exit lots per day or week). Typically there are different camps with substantially different views on how to make this decision. For example, manufacturing is looking to maximize output while the business team is concerned about the lots for key clients. As we will see in a future section, what looks easy on the surface gets complicated quickly.

4. Central Planning Basics and Challenges

This section covers what is often called supply chain management. This term is in fact unsatisfactory because it is really a demand supply network that can be viewed as a continuous matching process between assets and demand.

4.1 Overview of Best-Can-Do Enterprise Wide Central Planning Engines

The best-can-do (BCD) enterprise wide detailed central planning engine (CPE) is the control point for the flow of material or product within an organization and focuses on how to best meet prioritized demand without violating temporal, asset (WIP and inventory), or capacity constraints. A CPE application minimizes prioritized demand tardiness and some aspects of cost, establishing a projected supply and synchronized targets for each element of the supply chain.

The core of the CPE process is matching assets with demand, which refers to aligning assets with demand in an intelligent manner to best meet demands. The alignment or match occurs across multiple facilities within the boundaries established by the manufacturing specifications, process flows, and business policies.

Assets include, but are not limited to, starts (manufacturing releases), work in progress (WIP), inventory, purchases, and capacity (manufacturing equipment and manpower). Demands include, but are not limited to, firm orders, forecasted orders, and inventory buffer.

The matching must take into account manufacturing/production specifications and business guidelines. Manufacturing specifications and process flows include, but are not limited to, build options, bill of material (BOM), yields, cycle times, anticipated date on which a unit of

WIP will complete a certain stage of manufacturing (called a receipt date), capacity consumed, substitutability of one part for another (substitution), the determination of the actual part type after testing (called binning or sorting), and shipping times.

Business guidelines include, but are not limited to, frozen zones (no change can be made on supplies requested), demand priorities, priority tradeoffs, preferred suppliers, and inventory policy. Many of the manufacturing specification and business guideline values will often change during the planning horizon (time effective).

The creation of a CPE plan requires a solver (sometimes referred to as a model or an engine) with the following core features:

1. Method(s) to represent the (potential) material flows in production, business policies, constraints, demand priorities, current locations of asset, etc., and relate all this information to exit demand

2. Capture asset quantities and manufacturing specifications (parameters)

3. Search mechanism(s) to generate a balanced supply chain plan, relate the outcome to demand, and modify the plan to improve the match

4. Display and explain the results of the best-can-do match

The first task of any "best-can-do" CPE is to "flow material" and maintain a "feasible material flow" (see Graves et al 1995 for a review of material flow control mechanisms). Simply put, the CPE must decide a sequence of manufacturing starts to produce finished goods, and for each start the CPE places into the plan, the required component parts and capacity must be available and the manufacturing activity must be permitted on that day (e.g., it is not a shutdown day). For example, in Figure 4-1 if the CPE decides to manufacture 10 cards on day 10 to be completed on day 14 to meet a customer's demand, then on day 10 it must have 20 modules and sufficient tool/equipment capacity for the manufacturing process. Typically, the CPE handles this requirement either implicitly with material balance equations or explicitly with explosion and implosion.

Explosion and implosion are the core processes of the CPE which either move work units (WIP or starts) forward (implosion) to project completed parts or backward (explosion) to determine starts required across the bill-of-material (BOM) supply chain following the appropriate manufacturing data such as cycle time, yield, capacity, and product structure. We typically use implosion to estimate what finished goods will be available to meet demand and explosion to estimate what starts are needed at what due dates to ensure meeting the existing demand on time. Explosion and implosion are analogous to backward and forward scheduling in the finite scheduling literature.

To review implosion and explosion, consider Figure 4-1 again which represents a simple production flow. The first manufacturing activity is the production of Wafer_ 2. This manufacturing activity has a cycle time of 60 days, i.e., it takes on average 60 days to take a raw wafer and create a completed wafer with the part ID Wafer_2. The second activity is device production. Creating one unit of Device_2 requires 3 days and consumes 1/200th unit of Wafer_2. Module_2 consumes one unit of Device_2 and takes 8 days to produce. Finally, Card_2 consumes two units of Module_2 and takes 4 days to produce.

Referencing Figure 4-1, implosion can be illustrated with the following example. Manufacturing estimates that four units of Device_2 will be available or completed on day 10. This is called a projected receipt. If manufacturing immediately uses these four units to produce Module_2, on day 18 (10 + "Module_2 cycle time" = 10+8 = 18) four units of Module_2 will be completed. Continuing the projection process, the four units of Module_2 are immediately used to create two units of Card_2, which will be available on day 22 (18 + "Card_2 cycle time" = 18+4 = 22). The implosion process enables manufacturing to estimate the future supply of finished goods.

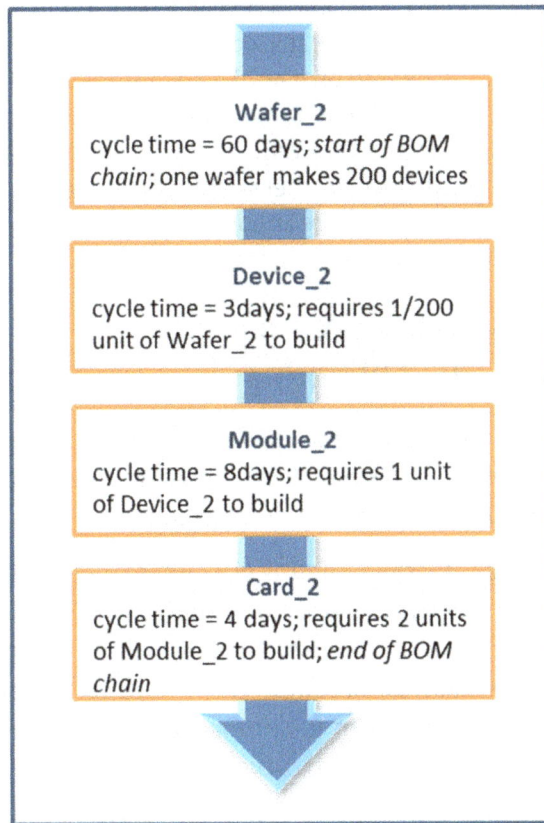

Wafer_2
cycle time = 60 days; *start of BOM chain*; one wafer makes 200 devices

Device_2
cycle time = 3days; requires 1/200 unit of Wafer_2 to build

Module_2
cycle time = 8days; requires 1 unit of Device_2 to build

Card_2
cycle time = 4 days; requires 2 units of Module_2 to build; *end of BOM chain*

Figure 4-1: Simple Flow for Production of Semiconductor Based Package Parts

Again referencing Figure 4-1, explosion can be illustrated with an example. To meet demand for one unit of Card_2 on day 20, the plant must have two (completed) units of Module_2 available on day 16 (20 − "Card_2 cycle time" = 20 − 4 = 16). This generates an exploded demand of two units of Module_2 with a due date of day 16. To continue the explosion process, to produce the two units of Module_2, the plant must have two units of Device_2 available on day 8 (16 – "Module_2 cycle time" = 16 – 8 = 8). Next, the device demand is exploded creating a demand for 2/200th units of Wafer_2 on day 5 (= 8 – 3). This exploded information creates the guidelines for manufacturing to meet existing demand. For example, the device department must start production of two units of Device_2 no later than day 5 to meet the demand for one unit of Card_2 on day 20. Since the cycle time to produce Wafer_2 is 60 days, it needs to have one already in production and close to completion.

Within the explosion and implosion process is a method called "demand pegging." This method links each allocation of an asset or creation of a start with either a specific exit demand, or, at a minimum, the demand class or priority (relative importance of demand) associated with the exit demand being supported.

Using the explosion example described above, if the exit demand for one unit of Card_2 on day 20 has a demand class of 3, each exploded demand will carry that demand class with it. Therefore, the units of Module_2 that are started on day 8 will have a demand class of 3. Similarly, if 3 units of Card_2 are desired on day 20 for a customer with demand class 5, then 6 units of Module_2 to be started on day 8 will also have a demand class of 5. The total required starts picture on day 8 is 8 (2+6), with 2 units with demand class 3 and 6 units with demand class 5. If by chance, there is only enough capacity on day 8 to start 2 units of Module_2, they will be allocated to the more important demand (demand class 3).

4.2 Challenges of Matching Assets with Demand in Central Planning

"The great 20th century revelation that complex systems can be generated by the relationships among simple components" [30] applies to supply chain planning (and almost all aspects of planning, scheduling, and dispatch).

Although simply creating a feasible central plan which maintains material balance, observes date effectivity, obeys business rules, captures existing WIP and inventory, and does a rough job at meeting demand is by itself challenging, it is no longer sufficient for a firm to remain competitive. The failure to "create a more accurate assessment of supply" forces the firm to compensate with slack (Galbraith 1973) or inefficiencies that leaves it at a competitive disadvantage.

The purpose of this section is to identify and describe some of the key challenges a CPE must handle to provide an accurate assessment of supply. By "handle," we mean provide a mathematical representation of the individual characteristics, the system or relationships, and a method to search for an intelligent, if not provably, optimal solution. In the following paragraphs, we describe the challenges of demand classification; simple binning; complex binning, substitution, and alternative bill of materials (BSA); lot sizing; sourcing; fair share; and commit date versus request date.

Leachman [52], Kempf [50], and Denton [14] also provide excellent reviews of the complexities/challenges within the production of semiconductor based parts. Leachman focuses on binning and demand class; Kempf develops and explains the stochastic challenge (uncertainty in demand and supply); and Denton deals with demand class and lot sizing in detail.

4.2.1 Allocating Resources Based on Demand Class

A fundamental decision found in most aspects of planning and scheduling is deciding which demand "gets to go first"? That is, when more than one demand needs either capacity (a perishable asset) or a component part (typically viewed as a non-perishable asset) and there is not sufficient supply to meet all immediate demand, the question is which demands get the asset and which have to wait. The CPE needs to allocate the asset based on the relative importance of the demand (indicated by demand class or priority) and the impact on delivering the finished goods on time.

Figure 4-2 shows a simple example of allocating supply of part (non-perishable asset) to meet demand. Module_1 and Module_2 are both made from Device_12 with cycle time 10 and 4 days, respectively. The demand for Module_1 is 10 units on day 10 and 15 units on day 12. The demand for Module_2 is 8 units on day 5 and 2 units on day 6. The key decision for the CPE is how to best allocate supply from Device_12 to the two modules (represented by the two boxes titled "Supply Amount" in Figure 4-2).

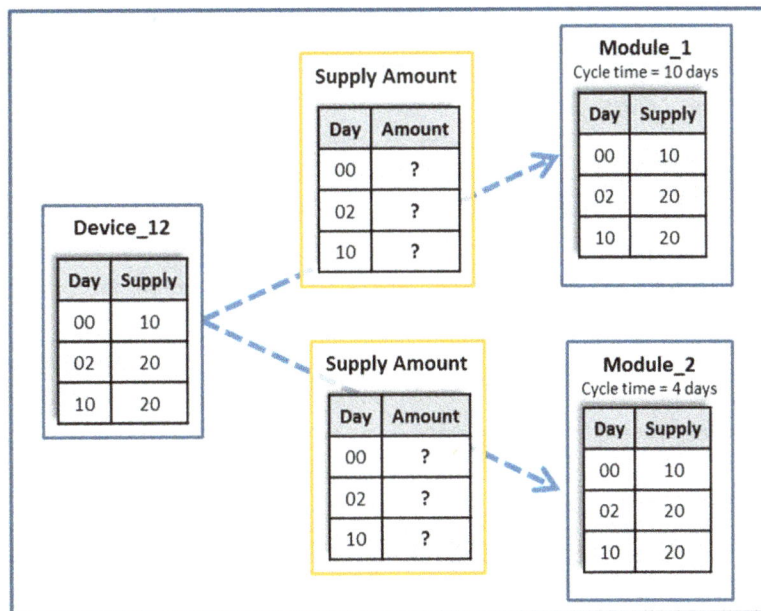

Figure 4-2 Example of Non-perishable Asset Allocation

One solution for the situation in Figure 4-2 might be: (1) immediately allocate 8 of the 10 units of Device_12 on hand to meet demand A (8 units of Module_2 on day 5) 1 day early on day 4 (=0+4); (2) immediately allocate the remaining 2 units of Device_12 on hand to meet demand B (2 units of Module_2 on day 6) 2 days early on day 4; (3) on day 2, allocate 10 units of the projected supply of 30 units of Device_12 to demand C (10 units of Module_1 on day 10) 2 days late on day 12 (=2+10); (4) on day 2, allocate 15 units of the projected supply of 30 units of Device_12 to demand D (15 units of Module_1 on day 12) on time (12=2+10). The score card for this solution is: demand A early; demand B early; demand C late by 2 days; and demand D on time. Table 4.1 summarizes this solution:

Demand ID	Type	Commitment		Actual Delivery		Delta schedule
		Date	Quantity	Date	Quantity	
A	Module_2	05	8	04	8	1
B	Module_2	06	2	04	2	2
C	Module_1	10	10	12	10	-2
D	Module_1	12	15	12	15	0

Table 4.1: Results of Solution 1

A second option could be: (1) 10 units of Device_12 are allocated to Module_1 on day 0 to cover demand C; (2) 15 units of Device_12 are allocated to Module_1 on day 2; and (3) 10 units of Device_12 are allocated to Module_2 on day 2. The score card for this solution is: demand A is met 1 day late, demand B is met on time, demand C is met on time, and demand D is also met on time. Table 4.2 summarizes this solution:

Demand ID	Type	Commitment		actual delivery		Delta schedule
		date	Quantity	date	quantity	
A	Module_2	05	8	06	8	-1
B	Module_2	06	2	06	2	0
C	Module_1	10	10	10	10	0
D	Module_1	12	15	12	15	0

Table 4.2: Results of Solution 2

Which is better? If the four module demands are exit demands, the answer depends only on the relative importance of each demand and the business policy on "sharing the pain" if a demand cannot be met on time. If demand A is demand class 1 (the lower the value, the more important the demand) and demands C & D are demand class 3, the first solution is the logical choice. If the demand classes are reversed, the second solution is the logical choice.

What if all of the demands have the same demand class? Do we go with solution two since demand A is just one day late? Do we meet part of demands C and A on time and the other part is late (can we split the order)? Or do we meet all of demand C and part of demand A on time? This typically depends on the business policy of the enterprise.

If the four module demands are not exit demands, then in addition to the demand class, we need to also assess whether meeting the module demand on time ensures meeting the exit demand on time. For example, if demand A goes into the exit demand for CARD01 for the XYZ customer and the board (modules go on boards to make a card) required is two weeks late, there is no point in worrying about meeting module demand A on time.

The CPE needs three core functions to handle this type of decision:

1. Ability to "peg" or pass down the priority of the exit demand to all intermediate demands

2. A method to pass in and represent the business policy

3. Mechanisms to search for options and assess the tradeoffs relative to overall efficiency and current business policy

Figure 4-3 identifies the basics for allocating limited capacity.

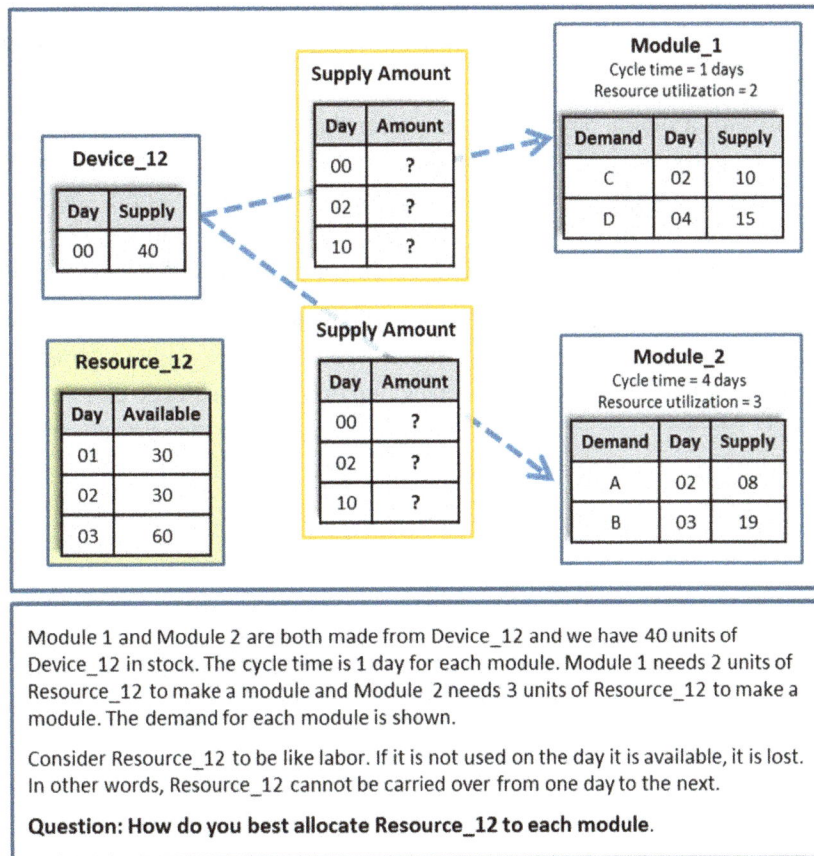

Module 1 and Module 2 are both made from Device_12 and we have 40 units of Device_12 in stock. The cycle time is 1 day for each module. Module 1 needs 2 units of Resource_12 to make a module and Module 2 needs 3 units of Resource_12 to make a module. The demand for each module is shown.

Consider Resource_12 to be like labor. If it is not used on the day it is available, it is lost. In other words, Resource_12 cannot be carried over from one day to the next.

Question: How do you best allocate Resource_12 to each module.

Figure 4-3: Allocating Limited resources

4.2.2 Simple Binning (or Sorting) with Downgrade Substitution

Simple binning with downgrade substitution refers to the classification of a part into one of a set of mutually exclusive and exhaustive categories based on some key performance factor. The classification is done after completing a set of manufacturing processes.

Quite often, parts with a higher performance can substitute for parts with a lower performance if necessary. This form of material substitution is generally called downgrade substitution, and occurs when there is a shortage of the lower performance part accompanied by an excess of the higher performance part.

In semiconductor manufacturing, the most common (but not the only) location for binning is the completion of wafer fabrication where hundreds of devices are cut from a single finished silicon wafer. Because of the random variation in the wafer fabrication process, working conditions for the devices on the same wafer vary. Therefore, before proceeding any further in the manufacturing process, devices have to be classified into different categories (each having a unique part number) after being cut from the finished wafer (typically today, the actual testing is done on the wafer). Clock speed, for example, is usually among a number of key performance factors to be tested for each device. Besides electronics, downgrade substitution can also be seen in such industries as consumer goods (bicycles) and building materials (grades of wood).

Figure 4-4 illustrates a typical binning scenario after the water fabrication process, where 50 percent of the time, the device is tested to have a "grade A" or top performance; 30 percent of the time, it has a "grade B" or medium performance; and 20 percent of the time, it has a "grade C" or low performance. These percentages are referred as binning percentages and can generally be observed in semiconductor manufacturing as a result of binning testing.

In Figure 4-4, the dotted arrows indicate that "grade A" devices (Device_1) can substitute for both "grade B" devices (Device_2) and "grade C" devices (Device_3), and "grade B" devices can substitute for "grade C" devices. The binning percentages can and do change over time.

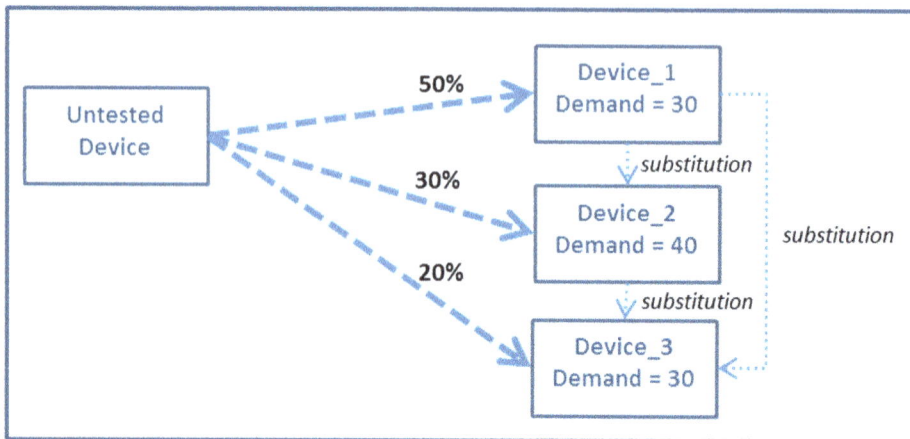

Figure 4-4: Simple Binning with Downgrade Substitution after Wafer Fabrication

The challenge is to make an optimal use of co-products and substitution to avoid overstating the number of wafer starts required to meet demand. If the demand is 30 for Device_1, 40 for Device_2, and 30 for Device_3 (Figure 4-4), the challenge is to determine the minimum number of wafers/devices that must be produced to meet all three demands. One simple rule is the maximum of the quantity required for each device divided by its binning percentage. Continuing the same example, we would need 150 devices, which equals maximum (30/0.50, 40/0.30, 30/0.20) = maximum (60, 133, 150). As shown in Figure 4-5, such a rule will leave an excess inventory of 50 devices, with 45 contributed by Device_1 and 5 contributed by Device_2. If we optimally account for co-products and substitutions, the minimum number of devices required to meet all three demands is 100—testing 100 devices creates 50 of Device_1, 30 of Device_2, and 20 of Device_3 (Figure 4-6). The extra 20 of Device_1 are used to cover the shortfall of 10 of both Device_2 and Device_3.

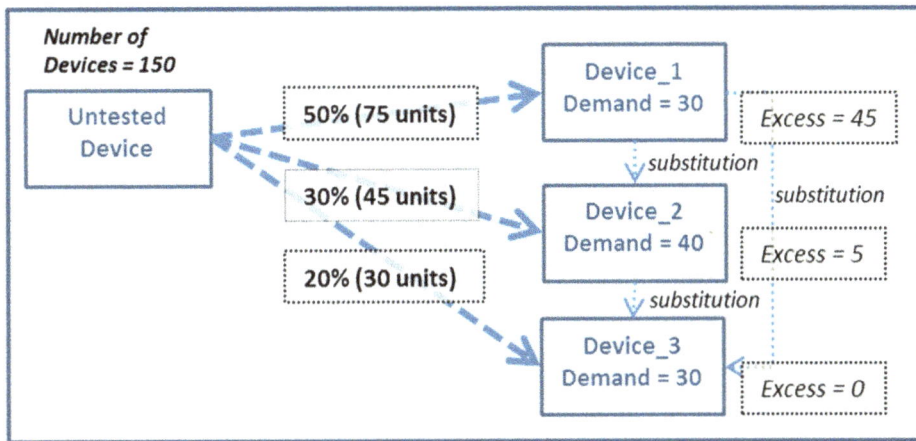

Figure 4-5: Maximum Quantity of Starts Leaves Excess Inventory of 50 Devices

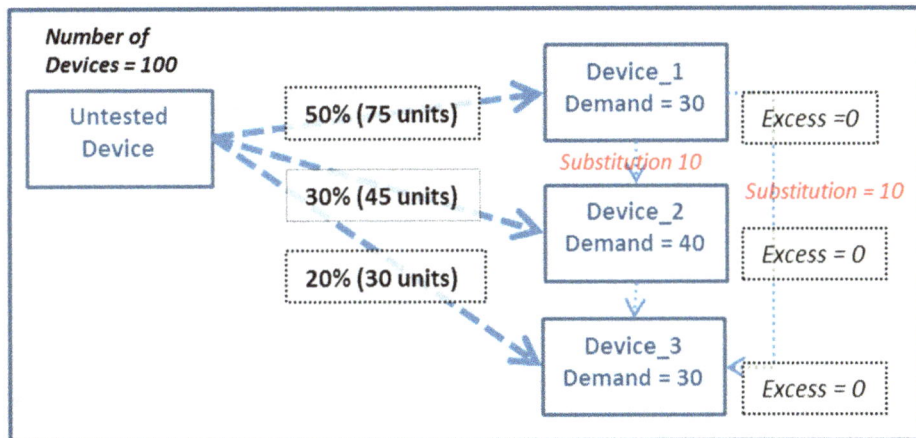

Figure 4-6: Optimal Number of Devices Meets All Demands and Leaves No Inventory

Other factors complicate calculating the minimum number of starts required to meet demand in simple binning production structures. These include: demands for devices that spread throughout the planning horizon, existing inventory, projected WIP completion, and binning percentages and allowable substitutions that change during the planning horizon (date effectivity). Additionally, the central planning engine must locate and isolate these binning situations in a large, complex demand-supply network, as well as maintain full traceability and handle demand priorities (Figure 4-7).

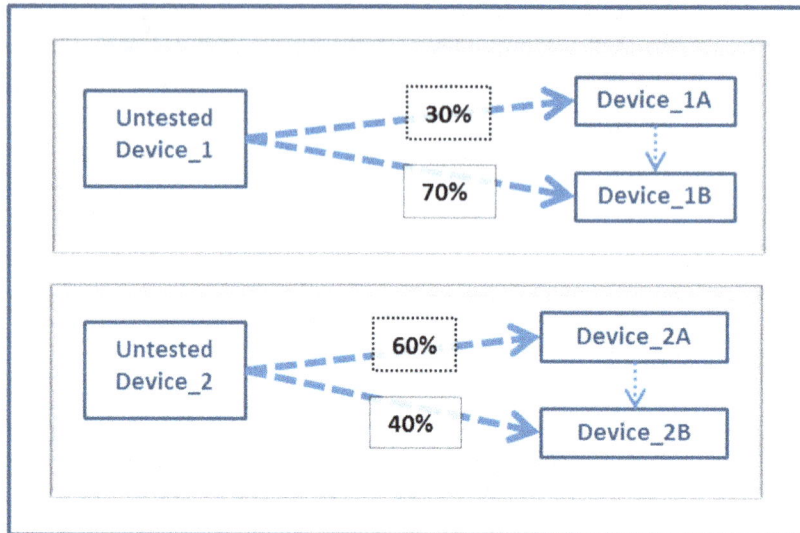

Figure 4-7: Two Binning Situations That Would Be Solved by Separate LPs

4.2.3 Alternative Bill of Materials and Complex Binning

Within the production of modules, an increasingly common manufacturing characteristic is alternative BOM (bill of material) structures, general substitution, and complex binning. In its simplest form, alternative BOM means two or more manufacturing processes are available to produce the same PN (part number). For example, Figure 4-8 illustrates a scenario in which process P1 and P2 can both be used to produce Module_9. If P1 is selected, the process will consume Device_8A; if P2 is selected, the process will consume Device_8B.

Complex binning refers to a situation where one binning activity immediately invokes another, or substitutions are permitted across binning activities.

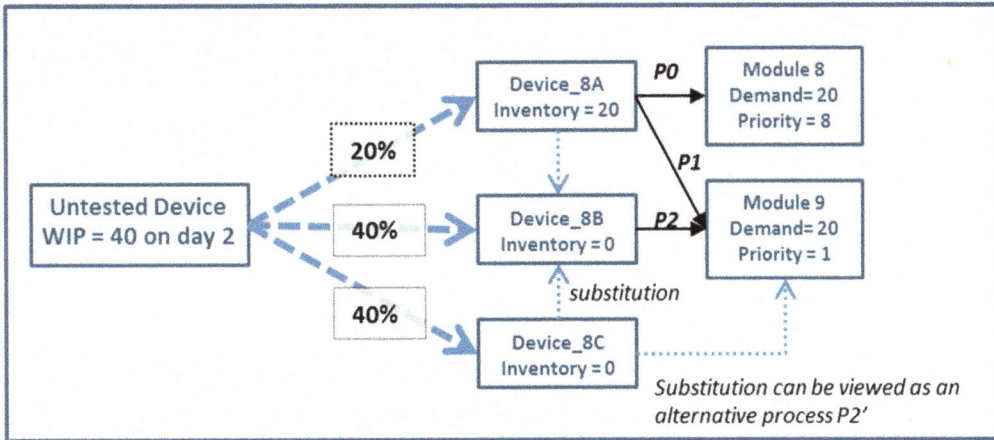

Figure 4-8: Illustration of Alternative BOM Structure with Substitution

To get a sense for the decision challenges created by alternative BOMs, we will use Figure 4-8 to look at just the "explosion" question within the CPE. Two processes (P1 and P2) can be used to build Module_9 where P1 consumes Device_8A and P2 consumes Device_8B. Additionally, Device_8C can generally be substituted for Device_8B, which means if Device_8B is not available and Device_8C is available, process P2 can use Device_8C to make Module_9. Conceptually, this general substitution can be viewed as the third alternative BOM option (call it P2-prime or P2').

The explosion engine must determine how to explode demand for Module_9 and Module_8 back to the device level. Should it be half to P1 and half to P2, 2/3 to P1 and 1/3 to P2, or all to P1? Should P2' be considered? The objective is to divide the demand for Module_9 across P1 and P2 (and perhaps P2') to best use the existing inventory and WIP, minimize new starts, and meet other relevant guidelines such as sharing percentages and capacity. Determining the best result requires an extensive search through the entire BOM structure.

In Figure 4-8, assume the priority for demand for Module_9 (which is 1) is higher than that for Module_8, and twenty units are ordered for both modules. A quick search reveals that twenty units of Device_8A are in inventory, which can be used to make Module_9 with process P1 or Module_8 with process P0.

There is no current inventory for Device_8B or Device_8C to make Module_9 using process P2 or P2' (the substitution). Most heuristic based search engines that guide the explosion through alternative BOM structures would explode the 20 units of demand for Module_9 down the P1 process (or leg). This would consume all of the 20 units in inventory for Device_8A and leave the demand for Module_8 totally uncovered; assuming building every unit of module would consume one unit of device (Figure 4-9).

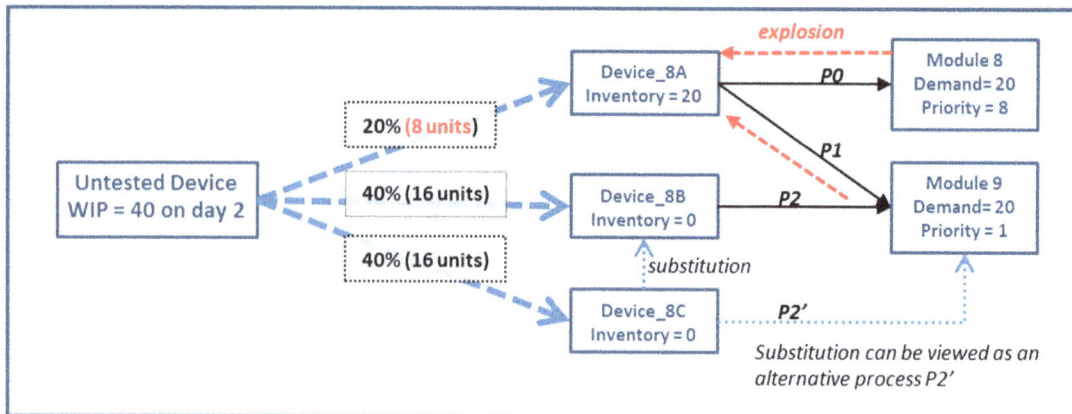

Figure 4-9: Awareness of Asset Availability Affects New Starts

To meet the demand for Module_8, a planning engine would require the demand-supply network to produce 100 new untested devices if the engine was not aware of the expected projection of the 40 untested devices (which could be tested into 8 units of Device_8A (=40*0.20), 16 units of Device_8B (=40*0.40), and 16 units of Device_8C (=40*0.40)). On the other hand, if a planning engine is aware of this information, it would only require 60 new untested devices since there is a projection of 8 Device_8A (60 = (20-8)/0.20 = 12/0.20).

However, a broader search would uncover all available options at untested device and avoid the conflict for Device_8A between Module_8 and Module_9. There are 20 units of demand for Module_8 and it can only be made from Device_8A.

So the question is: are there other options to meet the demand for Module_9? There are 40 units of projected WIP at the untested device and after binning, 8 will become Device_8A, 16 will become Device_8B, and 16 will become Device_8C. Since Device_8C can be substituted for Device_8B, there are actually 32 (16+16) future devices that can be used to produce Module_9 (but not Module_8), which is more than enough to meet the 20 units of demand for Module_9. So it is probably not optimal to explode the demand for Module_9 down the P1 leg.

As illustrated in Figure 4-10, both demands can be met without using any new untested devices:

1. Assign the 20 units of inventory for Device_8A to be used to produce 20 units of Module_8 (blue dotted arrow)

2. Explode the 20 units of demand for Module_9 into a need for 20 units of Device_8B (red dotted arrow). Net this need against the 16 projected Device_8B, resulting in a need for 4 new Device_8B

3. Use 4 out of the 16 projected Device_8C to meet the need for 4 new Device_8B

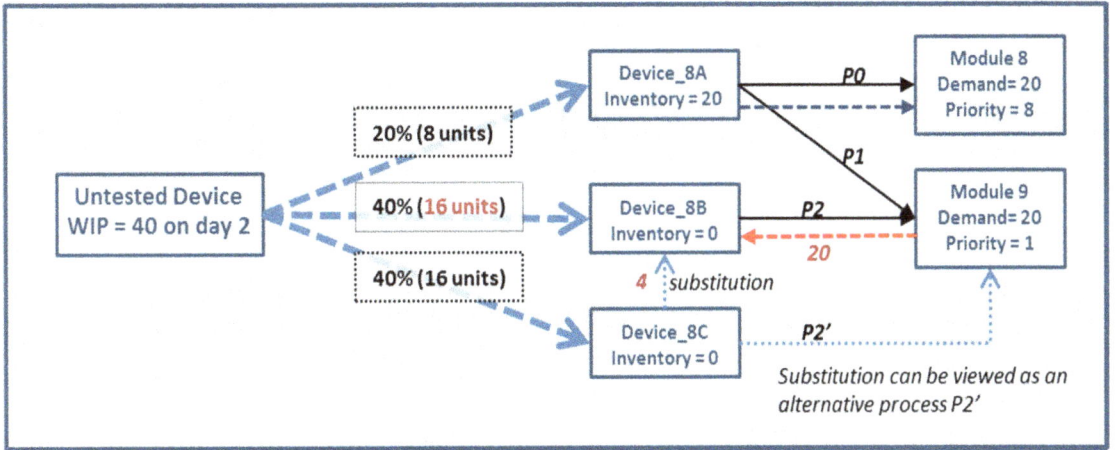

Figure 4-10: A Smarter Option—Meeting Both Demands without Using New Devices

Figure 4-11a and Figure 4-11b illustrates how the complexity can grow quickly and Figure 4-12 illustrates potential opportunities for parallelization through dynamic partitioning (indicated by different colors).

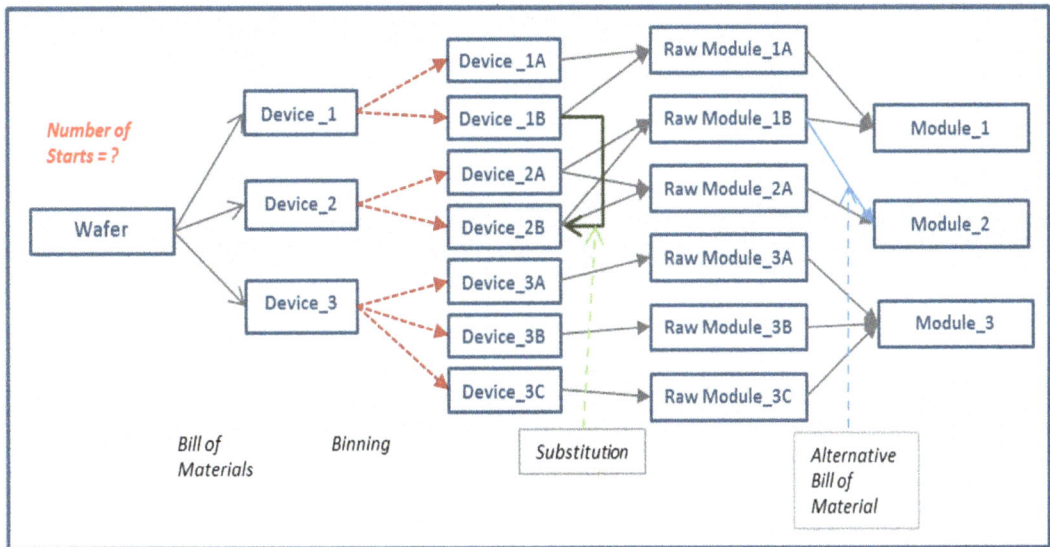

Figure 4-11a: Complexities across the Bill of Material Structure

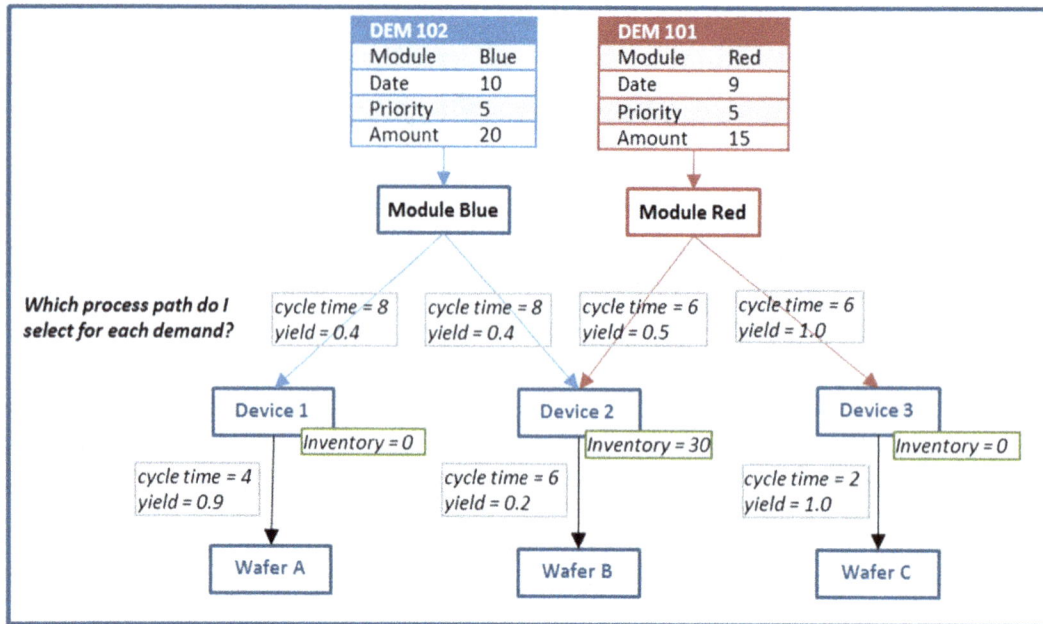

Figure 4-11b: Alternative Paths to Meet Demands that Share Common Assets

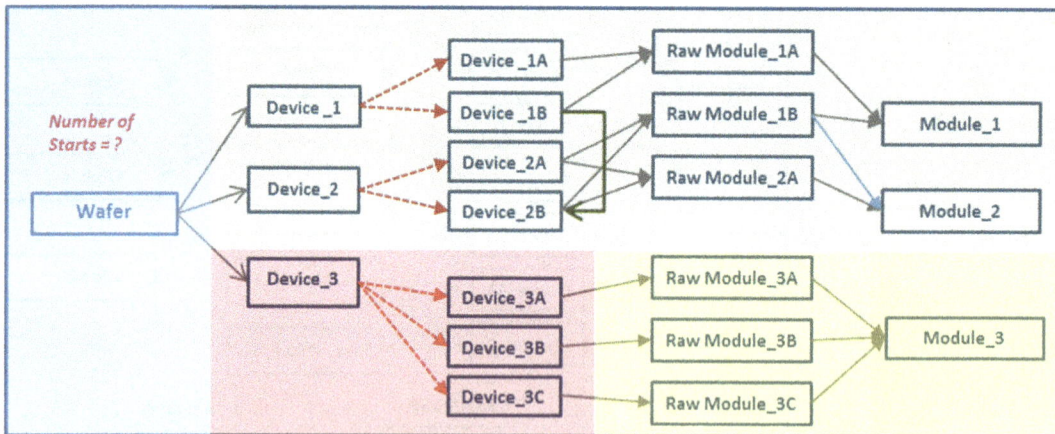

Figure 4-12: Opportunities for Parallelization through Natural Partitions

Increasingly, the CPE and the planner have to handle all three complexities: alternative BOM, general substitution, and complex binning simultaneously.

4.2.4 Lot Sizing

Lot sizing refers to a core manufacturing characteristic that the number of units in each lot of activity can have a significant impact on productivity. This preference/requirement is typically described by minimum lot size, maximum lot size, and multiples. Probably the easiest and most common example is minimum lot sizing.

Within a manufacturing facility, the same tool set is used to process a variety of products. For example, the same testing equipment is used to investigate and sort a wide variety of modules. However, there is a substantial "setup cost" when the tool set switches from one part to another. Therefore, the manufacturing unit wants to get a "return" on this setup cost by processing a minimum quantity of each part. The manufacturing unit does not want to do 3 of part A, then 5 of part B, then 2 of Part C, etc.; rather, it wants to do 30,000 of part A, then 5,000 of part B, then 20,000 of part C, etc.

This requirement creates a set of challenges for the CPE. First, the engine has to accommodate that in some cases the manufacturing release number can only come in discrete quantities (challenge for linear programming). Second, the engine needs a mechanism to keep track that the same lot will be accommodating different demand priorities. Third, not all lot sizing requirements are firm: the manufacturing facility may prefer to process 30,000 of part A instead of 3, but it can do just 3 if it is important enough.

4.2.5 Sourcing

If an enterprise has more than one supply location which provides certain parts, it typically wants to maintain some type of balance between the workload on each facility.

This "balance" is typically described with a reasonably complex set of business rules. Often, these rules arise from complex contractual obligations to suppliers that vary over time and business conditions.

Quite often, a contract specifies a minimum and maximum purchase quantity over a given period of time. The CPE is then required to plan manufacturing in such a way that the minimum and maximum purchase requirements are met.

4.2.6 Fair Share

Fair share refers to sharing limited supply or capacity among a set of equally important customers, as opposed to filling their orders in some random fashion. For example, A and B are the firm's most important customers, and they both have just requested 100,000 units of the same part to be delivered on the same day (200,000 units in total). Suppose the available supply is only 140,000 units, 60,000 short of the combined demand quantity. If no concern is given to the final delivery quantity, the CPE may generate a supply chain plan which ships 100,000 units to A (i.e., A's order 100% filled) and 40,000 units to B (i.e., B's order only 40% filled) despite the fact that they are equally important. With fair share, on the other hand, both customers will have 70% of their order delivered (and share the same degree of "pain"). The importance that all customers with the same level of importance to the company receive the same level of service is quite obvious. Planning for this requirement creates an additional challenge for the CPE.

4.2.7 Customer Request versus Customer Commit

Customer request versus customer commit (date, quantity, priority) is emerging as a key complexity directly impacting customer satisfaction. Typically, a customer requests a certain number of part(s) on a certain date and this is the customer request date and request quantity. Next, the enterprise reviews the request and responds with a date and quantity to which it will commit. This response becomes the commit date and commit quantity. The customer and supplier may iterate a few times on date and quantity until a firm commitment date and quantity are established. The supplier often associates some priority with the firm demand. Historically, this is the only date, quantity, and priority the central planning engine has for each demand; without using the original request date and quantity information during its solution process. In current practice, at best some post processing activity occurs to identify "low hanging" opportunities to meet a few key request dates.

To start, let's consider the simple example illustrated in Figure 4-13 where there is a demand for 100 units of part A with a request date of 06/01 and a commit date of 07/01. Assume WIP (work-in-progress) exists projected to come to stock on 06/01. A two-pass solution could do the following: in the first pass, WIP is used to cover the 07/01 commit date and then netted out and become unavailable for the second pass. In the second pass, no un-netted asset remains to cover the 06/01 request date. The two-pass process did not pick up the possibility that WIP could cover the request date. A solution that carried both the commit and request date would recognize the WIP could be used to make the request date with the primary search process.

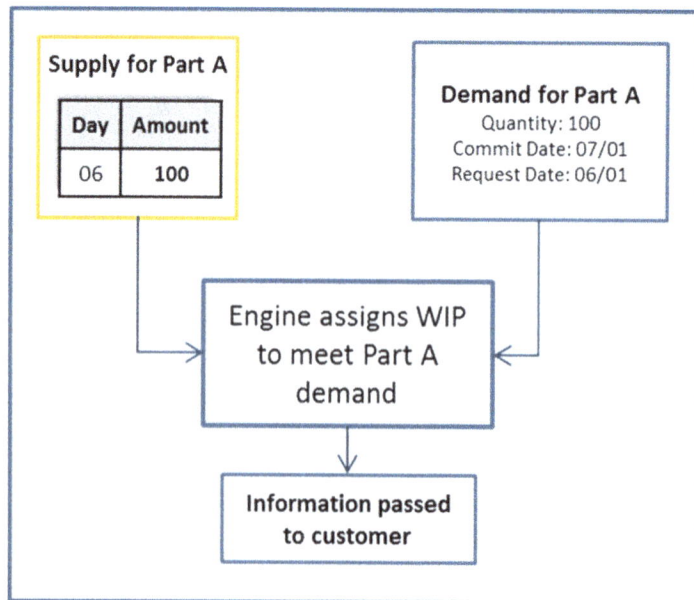

Figure 4-13: Simple Example of Handling Commit and Request Dates

A more complicated example is shown in Table 4.3. Here we have 5 demands for part XYZ and anticipated supplies arriving on days 2, 4, and 5. All demands have a commit date but only two (B and E) have a request date. To keep the example simple, we assume all demands have the same priority. The key business question is what date can we deliver the product to the customer! Can we meet the commit dates? Can we meet any of the request dates?

DEMANDS FOR PART XYZ			
ID	Commit Date	Request Date	Quantity
A	April 2	n/a	15
B	April 3	April 2	5
C	April 4	n/a	10
D	April 5	n/a	20
E	April 6	April 2	10
Total			60

ANTICIPATED SUPPLY FOR PART XYZ		
ID	Commit Date	Quantity
SUP01	April 2	15
SUP02	April 4	5
SUP03	April 5	10
Total		60

Table 4.3: Demand and Supply Information for Part XYZ

In Table 4.4, we show a typical solution focused only on the commit date—all the commit dates but none of the request dates are met. The on-time delivery (OTD) score for the commit date is 0, but the OTD score for the request date is -5.

Demands for Part XYZ				Allocated Supply to Meet Demand				On-Time Delivery	
Demand ID	Commit Date	Request Date	Quantity	Supply ID	Date	Amount	Date to Customer	For Commit	For Request
A	April 2	n/a	15	SUP01	April 2	15	April 2	0	n/a
B	April 3	April 2	5	SUP01	April 2	5	April 3	0	-1
C	April 4	n/a	10	SUP01	April 2	10	April 4	0	n/a
D	April 5	n/a	20	SUP03	April 5	20	April 5	0	n/a
E	April 6	April 2	10	SUP02	April 4	10	April 6	0	-4
Total			60			60		0	-5

Table 4.4: Solution Focused Only on Commit Date

Sometimes, a simple post processing routine is executed to identify "easy" opportunities to meet request date. In our example, the supply used to meet demand B has an anticipated supply date of 2, therefore the customer can have the supply on day 2. The results of a post processing routine are shown in Table 4.5. With this routine, we are able to meet the request date for order B but not the request date for order E.

Demands for Part XYZ				Allocated Supply to Meet Demand				On-Time Delivery	
Demand ID	Commit Date	Request Date	Quantity	Supply ID	Date	Amount	Date to Customer	For Commit	For Request
A	April 2	n/a	15	SUP01	April 2	15	April 2	0	n/a
B	April 3	April 2	5	SUP01	April 2	5	April 2	1	0
C	April 4	n/a	10	SUP01	April 2	10	April 4	0	n/a
D	April 5	n/a	20	SUP03	April 5	20	April 5	0	n/a
E	April 6	April 2	10	SUP02	April 4	10	April 6	0	-4
Total			60			60		1	-4

Table 4.5: Solution Obtained with a Simple Post Processing Routines

Table 4.6 shows the results of carrying both commit date and request date within the solution process. Both request dates (demands B and E) are met. This is accomplished by swapping SUP01 and SUP02 between demand C and E.

Demands for Part XYZ				Allocated Supply to Meet Demand				On-Time Delivery	
Demand ID	Commit Date	Request Date	Quantity	Supply ID	Date	Amount	Date to Customer	For Commit	For Request
A	April 2	n/a	15	SUP01	April 2	15	April 2	0	n/a
B	April 3	April 2	5	SUP01	April 2	5	April 2	1	0
C	April 4	n/a	10	SUP02	April 4	10	April 4	0	n/a
D	April 5	n/a	20	SUP03	April 5	20	April 5	0	n/a
E	April 6	April 2	10	SUP01	April 2	10	April2	4	0
Total			60			60		5	0

Table 4.6: Solution in Which Both Request Dates Are Met

The challenge is to carry both commit and request information (date, quantity, priority, etc.) intrinsically within the CPE and identify opportunities to come closer to customer requests (respecting their relative importance) without sacrificing any commitments which have been made. This requires carrying multiple dates and quantities throughout the explosion component of the CPE, identifying opportunities to assign assets to exploded demand to improve the posture of the exit demand, and being robust enough to handle such complexities as binning, substitution, and lot sizing.

4.2.8 Minimum Starts

Sometimes, a planner will want the CPE to ensure that a minimum number of starts occur at a certain point in the production process. For example, in Figure 4-1, the planner may want a minimum of 100 Module_2 per week over the next 20 weeks to smooth the manufacturing flow or to meet contractual agreements (e.g., a vendor may be willing to promise a shorter cycle time if it has a "smoothed" start plan). In these cases, the CPE must explode the exit demand back to the identified minimum start manufacturing activity, compare the minimum start value with the required value, then continue the explosion process. On the return trip (implosion), the CPE must also adapt for the relative priority of the minimum start.

4.2.9 Date Effective Parameters

"Date effective parameters" are a challenge that is often underestimated and underserved—the solution offered often does not do an efficient job at meeting this challenge, leaving a firm with last minute patching and manual intervention.

Simply, this challenge is that most "descriptive" elements of the demand-supply network, such as yields, cycle times, capacity available, capacity consumed, allowable substitutions, BOM, and so on, have a start date and an end date (date effective), and the CPE must recognize that these elements will change over time. For our discussion we will use cycle time.

Previously, we used Figure 4-1 as a simple production flow to explain explosion and implosion. In that example, the cycle time was fixed over time. Let's make the cycle time for Module_2 and Card_2 date effective (Figure 4-14). The cycle time to produce Module_2 is 8 days in duration from day 1 to day 10 of the planning horizon and 10 days from day 11 to day 25. This means that the production of Module_2 started on day 5 has a cycle time of 8 days and will complete on day 13 (5+8). But if the production starts on day 12, the cycle time becomes 10 days and the completion time will be day 22 (12+10). The cycle time for Card_2 is 4 days from day 1 to day 14 and then reduces to 2 days from day 15 to day 25. Therefore, the search engine in both explosion and implosion must account for these changing cycle times.

Now let's revisit our implosion example. Manufacturing estimates four units of Device_2 will be available on day 10. If manufacturing immediately uses these four units to produce Module_2, the cycle time will be 8 days and on day 18 (10+8) four units of Module_2 will be completed. Continuing the projection process, the four units of Module_2 are immediately used to create two units of Card_2 beginning on day 18 (therefore the cycle time is 2 days) which will be available on day 20 (18+2).

What if capacity was not available on day 10 to start the production of Module_2? If capacity became available on day 11, the cycle time for Module_2 would increase from 8 days to 10 days. Therefore, a one-day delay due to a capacity restriction would result in the completion of Card_2 being delayed 3 days to day 23 (11+10+2).

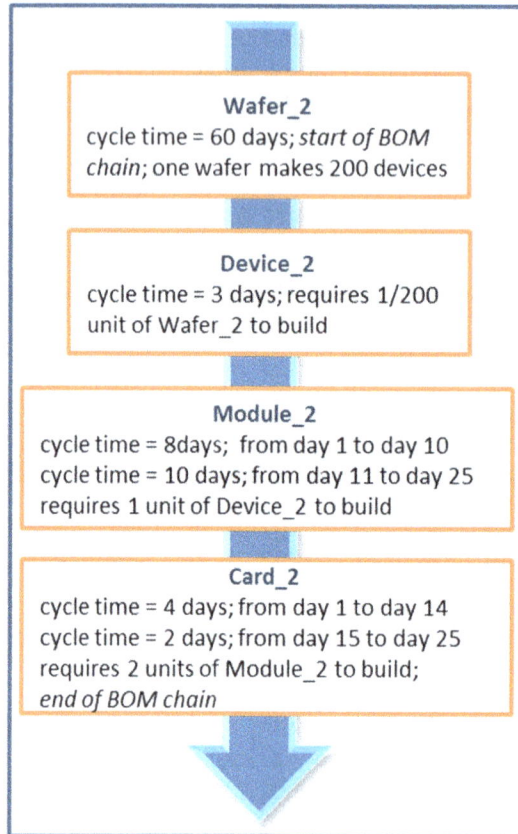

Figure 4-14: Flow for Production of Semiconductor Parts with Date Effective Cycle Times

Let's turn our attention to explosion. To meet demand for one unit of Card_2 on day 20, the plant must have two completed units of Module_2 available on day 18 (20 minus the cycle time for Card_2 on day 18 = 20-2 = 18). This generates an exploded demand of two units of Module_2 with a need date or due date of day 18.

When does the manufacturing facility need to start the production of two units of Module_2? When we had only one cycle time for the module, we simply subtracted that value (8) from the module need date. This would drive a Module_2 start on day 10 (18-8). Life is not that simple with date effectivity.

As a starting point, we first find the cycle time for Module_2 on day 18 which is 10 days. Remember 10 is the cycle time for the production of Module_2 that starts on day 18. But we do not want to start production on day 18; rather, we want production to complete (called to stock) on day 18. So we subtract 10 from 18 and get an initial starting date of 8. Does this work? We check the cycle time for the production of Module_2 on day 8 and find it is 8 days, meaning that if we start the production on day 8 the module will be completed on day 16, two days earlier than the time we need them.

We might just stop here, since we have found a feasible solution. However, another demand may need a start on day 16. We extend our search to determine if we can start later than day 8. Logically we might try day 10. The cycle time for the production of Module_2 on day 10 is still 8 days, so the modules will be completed on day 18 (10+8). Just in time!

This is not the only complexity associated with date effectivity. Two other common ones are aggregating cycle times and converting to time buckets. In cycle times, we might lump the production of Module_2 and Card_2 into one production activity (call it Mod_Crd_2). Before introducing date effectivity this was simple—just add the two cycle times and it becomes the cycle time for the new activity (8+4=12). Now we need to adjust for cycle time changes. In this case, the cycle time for Mod_Crd_2 is 12 days from day 1 to day 6, 10 days from day 7 to day 10, and 12 days again from day 11 to day 25. In time buckets, we need to convert the cycle time from daily information into number of buckets. For example, if the time bucket was 3 days in duration, we have 8+ buckets (25/3) and we need to restate each cycle time from days to buckets. Some of the buckets are located on a date effective boundary.

4.2.10 Illusion of Capacity

With very few exceptions the complexities of capacity and cycle time (and their interaction) in FABs has been ignored in the central planning process and engine (model). In fact these complexities are often, but not always [1, 82, 91], "understated" in detailed aggregate level FAB planning.

Typically in central planning FAB capacity is stated as a set of nested wafer start (or exit) limits at high levels of aggregation and cycle or lead time is stated as a fixed entity in a separate file. These two pieces of information are treated as fixed and unconnected inputs to central planning. The following table has an example of "aggregated nested start" limits (Table 4.7)

In this example the overall FAB limit is stated in terms of wafers per day and that each product is mapped to one or more limits. These methods allow the central planning model and process to start up to, but not over any limit to which products are mapped.

For example, a part mapped to Option Set W in Tech Group B in "Wiring group 1" may not exceed 100 wafers per day AND the sum of Option Sets W and X must be no more than 300. Therefore starting 100/day of 'W' limits the start rate of 'X' to no more than 200 in order to avoid the limit for Tech Group B, even though the limit of 'X' is stated at 300/day. In reality, within each of the current start-limit groupings there are several distinct products. Each has a unique path through the FAB and a distinct set of operations for each of the FAB toolsets.

FAB limits as a nested set of Start Limits			
Group	Time frame 1	Time frame 2	Time frame 3
Wiring Group 1	600	675	675
Technology Group A	400	425	450
Technology Group B	300	325	350
Option set W	100	100	100
Option Set X	300	300	300
Wiring Group 2	500	525	550
Technology Group D	350	350	375
Technology Group E	250	275	275
Option set Y	100	100	100
Option Set Z	200	200	200
Engineering and other starts	100	100	100
Total Fab Limit	1000	1100	1150

Table 4.7: FAB limits as a nested set of Start limits

With very few exceptions the twin complexities of FAB aggregate tool planning – deployment decisions and the operating curve – have been at best ignored and this a ripe area for improvement.

4.2.11 Other Technical Challenges

Other challenges include, but are not limited to, demand perishability, squaring sets, soft capacity constraints, alternative capacity, pre-emptive versus weighted priorities, splitting demand to match partial delays in supply, stability, express lots, delay assembly to test, dispatch lots, foundry contracts (Figure 4-15): service model; multiple exit demands along the same bill of material supply chain), risk based inventory policy, multi-supplier sourcing using inventory, WIP projection, rules governing purchase order change recommendations (allow for date change, quantity change, both or neither).

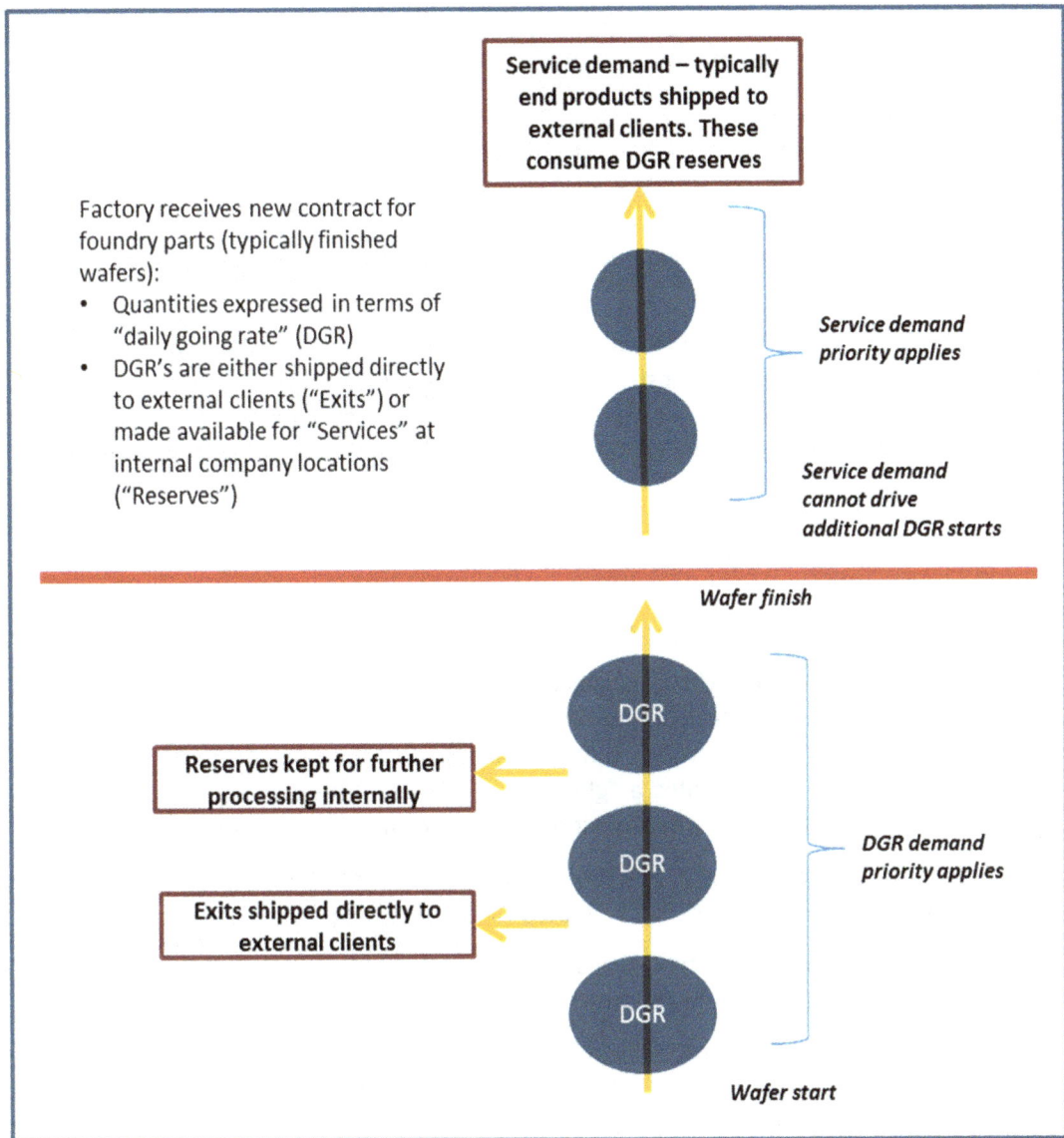

Figure 4-15: Service and DGR Demands for Foundry Contracts

4.2.12 Variability in the Demand-Supply Network

Uncertainty (in parameters, estimated supplies, projected demands, etc.) is no doubt another critical challenge, but we will only briefly touch on this topic in this section.

In the near term, uncertainty may force planning applications to take a conservative approach to risk. For example, if (a) the average binning percentage for fast parts is 30%, (b) it has a uniform variability of plus/minus 10%, and (c) there are 100 parts currently being tested, then the actual number of fast parts that we will get from this specific manufacturing activity ranges from 20 to 40. Although on average we will get 30 fast parts from the sorting operation, 50% of the time the actual number of fast parts will be less than 30.

Working with the average value is fine over a moderate or long time frame, but it is impractical if your time frame is only one day. One proposed method called cycle variation looks at taking a more conservative approach with high priority demands than with low priority ones. The downside of this method is that "lower priority demands" had a higher probability of the central planning engine determining they could be met on time—a small problem!

For long range planning, executives would prefer to understand the range of possible outcomes and their likelihood, instead of being given a point estimate. There is a range of work going on (Kempf [51]) with some involving stochastic optimization while others pulling wafer fabrication capacity planning [91] and still others exploring inventory policy [41].

4.3 Executing the Plan

4.3.1 Limitations of Current Approach

We refer to today's supply chain management central planning process as the "big bang" approach. An enterprise creates a centralized process and data representation of the firm at some point in time. Then, through some combination of automated models and manual processes, the enterprise creates a "global plan." Some firms execute a global planning process a few times each day; others do it once a month and take a week to create the plan.

In either case, the new plan arrives after a reasonable amount of effort and replaces the old plan in its entirety. A new "universe" is created via the "big bang!"

This approach requires solution times to be reasonable. Techniques include parallel processing, faster algorithms, faster computers, and stricter modeling assumptions.

As solution times have improved, some companies have taken opportunity to improve accuracy with smaller time buckets or units of granularity and with more robust representations of the demand-supply network.

Although centralized processes and models have increased organizational effectiveness, there are clear limits and we are rapidly reaching them.

Effective centralization refers to the ability to take into consideration essential aspects of the decision situation simultaneously and generate an optimal or at least a very good solution. To be effective, a central solution requires a synchronized current view of the entire decision landscape, the ability to handle complex tradeoffs, and a reasonably fast runtime performance.

At the start of the 20th century, physicists learned that we do not live in a clockwork universe [84]. The same can be said for supply chain modelers at the start of the 21st century.

Gaps exist and are often created by time lags, summarization, performance, triggers, and formulation. By triggers, we refer to the event that "wakes up" at a specific time during the day and runs the central solution. Once a week, once a day, or once every 3 days, the central solver executes, but the decision to execute is made without any knowledge or monitoring of events since the last execution.

Formulation gap refers to the inability to formulate key decision questions in such a manner that lends itself to a central solution as opposed to a sequence of negotiations or collaborations. The reality is, even if we could get the big bang models to finish execution in 5 minutes, an organization would run it once a day at best to match demand with supply and synchronize the enterprise. As the time between runs decreases (for example, from once a week to once a day), the following limitations become apparent:

1. Understanding and repairing the plan:

    1.1. Why does the plan give these results?

    1.2. What demands are not being met (alerts) and why?

    1.3. What actions can I take to improve the plan?

        1.3.1. Identify and book actions to improve the supply posture as it relates to demand.

        1.3.2. Less significant items such as inventory picture.

    1.4. Monitoring the repair actions as they are being executed.

2. The nature (quality) of the plan – no plan to plan continuity:

    2.1. No checks, filters, or alerts (CFAs) on demand information / signals.

    2.2. No CFAs on supply information / signals.

    2.3. No CFAs on changes in production specification or business policy.

    2.4. Each plan is built from scratch.

    2.5. No build-in dialogue with other key providers of input, such as

        2.5.1. Projected supply.

        2.5.2. Projected demand.

        2.5.3. Capacity estimation

Observe that both 1 & 2 require an incremental matching (planning) or net change engine

4.3.2 Living with the Plan

"Optimizing" the initial plan is just one part of "optimizing" the entire solution which includes robust estimates of capacity and cycle time, a true understanding of the demand patterns, the process of building and "promoting" a plan to record, etc.

A firm that wants to be responsive has to see the initial plan as the base and provide its planners the software tools to navigate and adapt the plan. It has to enable the planner to easily identify what set of enterprise activities are linked together and in what order to support an exit demand; try out alternatives including options to override capacity restrictions (which are just estimates); and "book" incremental modifications to the plan without "rerunning" the entire plan. It needs to provide its planners an environment that mimics making a small plan "manually" on large white board with the computational ability to manage large complex plans. That is answering the question – "how is this order being satisfied and what if I do something different?"

Simple Repair Actions Becomes Complex

The following example shows how a "simple" repair action can quickly become complex. In Table 4.8 we have two demands for the same part (P111): D001 with a due date of April 4, priority of 2, and quantity of 80 and D002 with a due date of April 5, priority of 1, and quantity of 100.

ID	Date	Priority	Quantity
D001	April 4	2	80
D002	April 5	1	100

Table 4.8: Demands for Part P111

In Table 4.9 we have two anticipated supplies for part P111: S00A and S00B with an anticipated delivery date of 4/4/2006 and 4/6/2006; the quantity is 100 units each.

ID	Date	Quantity
S00A	April 6	100
S00B	April 6	100

Table 4.9: Supplies for Part P111

Initially, we assign S00A to D002 (Figure 4-16) since D002 has a higher priority and the other supply S00B would arrive 1 day after the due date for D002. This assignment resulted in D001 being met 2 days late.

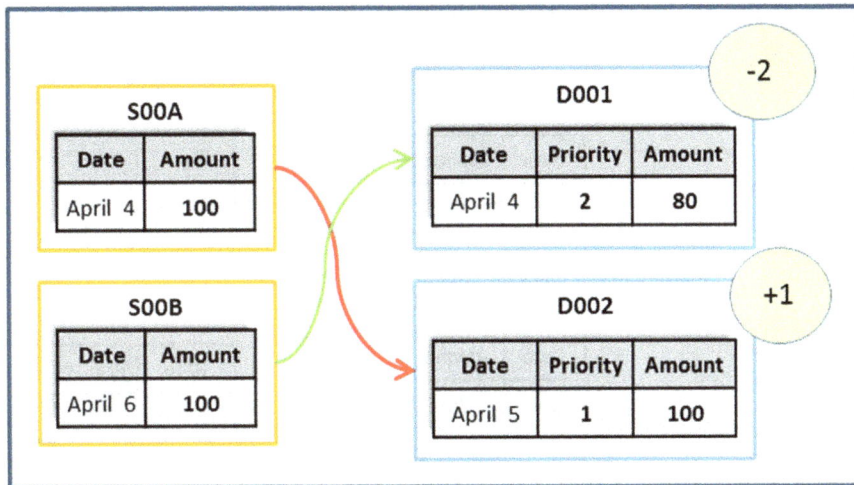

Figure 4-16: Initial Assignment of Supply to Demand for Part P111

Figure 4-17 demonstrates a simple repair action to meet demand D001 on time: the analyst requests that the S00B supply be completed two days earlier on 4/4/2006.

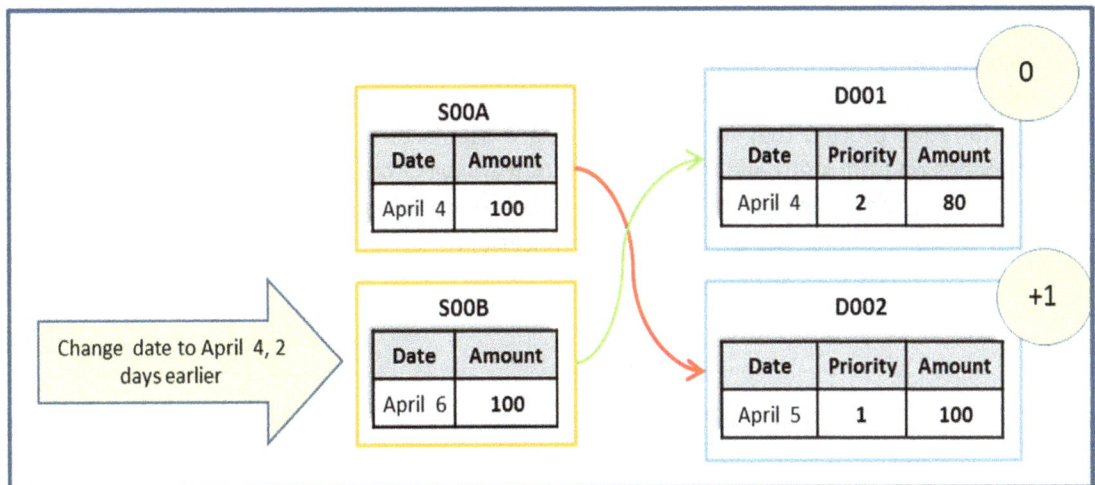

Figure 4-17: Simple Repair to Meet Demand D001 on Time

Figure 4-18 shows a smarter repair action. First, the analyst switches the assignment of supply to demand – supply S00A is assigned to D001 and supply S00B is assigned to D002. Then the analyst requests that supply S00B be expedited only 1 day to 4/5/2006 to meet D002 on time.

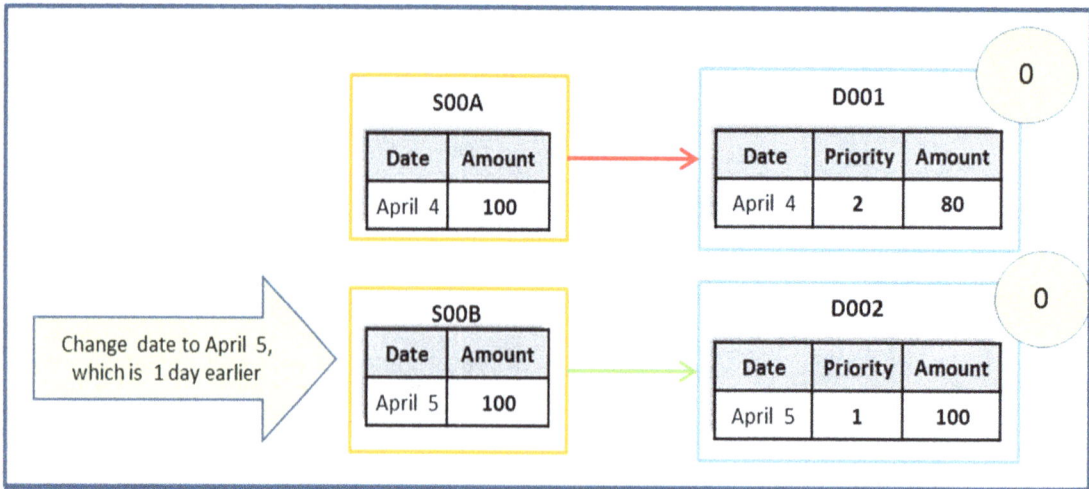

Figure 4-18: Smarter Repair Action to Meet Both Demands on Time

To take organizations to the next leaps in efficiency requires a substantive adjustment in approach. Just as 19th century physics had to adjust its equations and formulations for special relativity, general relativity, and quantization, supply chain solvers must learn to accommodate techniques from intelligent agents and sense and respond and learn to incorporate the concept of collaboration into their solution processes. Collaboration refers to an iterative process that focuses on finding a satisfactory solution. The next search step depends on prior steps and may involve back tracking. Often, the step involves negotiating a temporary change in a subset of the rules governing the game for a limited period of time, and typically contingency occurs to handle uncertainty.

5. Factory Planning Basics and Challenges

Historically factory and especially FAB planning has been disconnected from central planning. In fact often FAB planners know little about central planning and vice versa. FAB planners are concerned with yields, cycle times, and utilization, while central planners are concerned more about due-dates, promise dates, and stock availability.

5.1 Overview of Factory Planning

Factory planning, like central planning, is focused on matching assets with demand to determine what commitments can be made to the business and what opportunities to chase.

To support central planning, it has to provide the central planning engine with a statement of available capacity. In order to do this, at the very minimum, it must:

1. Determine which toolsets are going to be used and for which operation

2. Calculate the projected cycle time based on the factory utilization and the central planning engine's projected starts

3. Provide the central planning engine with an estimate of when the current WIP will be available for further processing

It is true that the interaction between factory planning and central planning is somewhat of a "chicken-and-egg" situation because the projected cycle time cannot be calculated without assuming some sort of factory load, and the required factory load cannot be calculated by the central planning engine without some idea of the cycle time.

In practice, the factory planning is done on a less frequent cycle and for a longer time horizon than central planning and generally tends to drive central planning.

5.1.1 Aggregate Factory Planning

Aggregate planning [1, 79, 87, 88] is focused on assessing the ability of the factory to meet the requirements of a central plan by looking to identify "broken" toolsets and creating the capacity inputs required by central planning model.

Requirements for the factory are typically stated as a "starts" profile and a lead (cycle) time commit for each part. For example, the factory has been asked to start 100 of Part A each day with a lead time of 20 days and start 50 of Part B with a lead time of 10 days.

Historically, the focus is typically to identify "broken" tool sets (groups of tools currently without sufficient capacity to meet this load (starts) and cycle time).

Once these tools sets are identified there are only two options:

1. Take actions to add capacity by reallocating tools sets to operations
2. Have central planning modify the load profile

Typically (but not always) this is a slow and reasonably static process compared to the more dynamic central planning process. It operates more like an MRP (which simply tells the central planner what needs to be done to achieve a given demand), then a "best can do" model (which searches a wide set of options to best meet prioritized demand without violating constraints).

That said, there is strong interest and need to upgrade this process to dynamically handle such factors as demand priority, cycle time versus load trade-off, and partially shared operations between tools.

The results of factory planning directly influence the capacity available to the central planning engine. Often the two processes are run iteratively to arrive at a reasonable overall solution.

Why is this different then central planning? Aggregate factory planning considers:

- The actual route (sequence of manufacturing actions to produce a finished part)
- The trade-off between cycle time and tool utilization / capacity available
- Partially Shared Operations (PSO) or Cascade which refers to different toolsets that overlap in the manufacturing operations they service, but the actions are not identical
- Deployment Variability
- Product mix transitions
- Reentrant flow

All but the first item is described in more detail in later sections and make up much of the core of the 'ongoing challenges".

5.1.2 Deployment or Near Term Tool Planning

Typically a tool can service many operations, but a factory will limit the number of operations it is "allowed" to service. Deployment refers to designating which operations each tool will be qualified to handle over the short term.

While the stated reason for restricting tools to certain operations is often that it improves speed and yields by focusing the tool on the operations that it is best suited to handle, there are in fact other practical benefits:

- It ensures that a tool remains qualified to handle a particular operation by regularly servicing this operation

- Restriction of tools to operations may be needed because of specific restrictions which may limit a tool from serving some operations based on the operations it can currently service

- Restriction reduces the work load on manufacturing engineering to maintain tool qualification

- Dispatching decisions are made easier because there are fewer options

5.1.3 WIP Projection

All central planning engines require the factory to project when the current WIP will be completed or reach a staging point. The typical process involves capturing the remaining steps (route) for the lot and estimating how long it will take to finish these steps.

5.2 Challenges and Opportunities for Aggregate Factory Planning

Over the past twenty years organizations have put significant energy into making smarter decisions in their enterprise wide central planning and "available to promise" process to improve responsiveness. This has resulted in more effective use of assets and more intelligent responses to customer needs and emerging opportunities.

However, firms have put limited energy into factory floor decisions and capacity planning, and almost none into a tighter coupling between factory and central planning. Much of the recent work to improve factory performance has attempted to implement core lean planning [10] concepts.

These focus on eliminating variability, establishing uniform flow (every part every interval), supermarket goods flow (kanbans), and eliminating due dates. Clearly every factory will run "better" with steady output and predictable lead times—however the real world always injects variability that sets the price of implementing such methods as reduced responsiveness and/or excess capacity.

The net result is that many factories still operate with the mind set: "establish a set of starts for the month; set a fixed schedule with target outs; measure actual outs versus target outs". For this approach to work, demand must be accurately forecasted over an extended period of time and uniformly spread across time; all lots travel at the same speed, tool sets operate with clockwork precision (never suffering "surprises"); and the flow of parts in the line (even with stable capacity) never creates "piles" or "gaps" due to the variations (for example batch versus single lot tools) intrinsic in the manufacturing process.

In today's world, accurate detailed forecasts of demand are an illusion; even the best factories have "tool set surprises" (breakdowns and excursions). Product mix introduces variability in speeds, and the competitive nature of the market precludes carrying excess capacity and insists on responsiveness.

Those demand supply networks that can get their factories engaged in responsiveness while recognizing the importance of "tools" and "output" will flourish; they will eliminate the variability that matters—a failure to deliver a part on its committed date and the inability to capture a market opportunity that could be handled with "intelligent" factory decisions.

5.2.1 Some Inconvenient Complications

In the traditional world of central planning, capacity is modeled with:

- simple linear equations where the two primary inputs are resource required and resource available (for every desk produced we need 4 legs, 1 top, and varnish)

- the cycle time or lead time is fixed and disconnected from capacity

- capacity allocation is usually focused on manufacturing releases, not work in progress

Factory planning features such as the relationship between utilization and cycle time, reentrant flow, partial sharing of operations between tools, and deployment decision are a level of complexity that is not considered in linear models.

The trade-off between capacity and cycle time

When variability exists either in arrivals or processing time, there is a trade-off between server (tools, people) utilization and the lead or cycle time to complete an activity.

All factory planners have observed that the cycle time does not increase linearly with utilization. As the utilization increases, variability plays a greater and greater role because there is less chance that a resource is available exactly when it is needed. This increases the waiting time between operations much like the customer line in front of a heavily utilized cashier at a supermarket.

The cycle time is often measured as a cycle time multiplier (CTM), where CTM equals total elapsed cycle time divided by raw process time (RPT). In the best case, the CTM is exactly 1 because all resources are available when they are needed. Typically the CTM is almost flat for low utilization levels, then spikes sharply upward when the utilization increases.

The point at which the CTM starts increasing sharply depends mainly on the processing time variability. This behavior can be modeled in many ways. We have found it convenient to use the formula:

$$CTM = 1 + offset + \alpha \left[\frac{util^M}{1 - util^M} \right] \quad \text{(Morrison and Martin [65])}$$

- **CTM** is the cycle time multiplier

- **util** is the tool utilization in the entity – facility, tool set,

- **offset** represents several of aspects of the process that generate wait time and cannot be eliminated. For example: travel time, hold hours, post processing hours relative to the processing time. A typical value for offset is 0.9. When offset is 0.9 this sets the minimum CTM at 1.9.

- **M** is the number of identical parallel machines or servers. Typically this value ranges from 1 to 4 (even when the number of tools exceeds 4)

- **α** represents the amount of variation in the system and controls how long the curve stays flat. The lower the value of **α,** the less variation and longer the curve stays flat. Common values for α range between 0.35 and 0.65.

Equivalently, the utilization can be expressed as:

$$Util = \left(\frac{CTM - (offset + 1)}{CTM - (offset + 1) + \alpha} \right)^{\frac{1}{m}}$$

Linking Capacity and Cycle Time

Assume that the product XYZ is processed 5 times by tool set AAA during its production route. Each time a widget goes through tool set AAA it is referred to as a "pass". In this case product XYZ has 5 passes on tool set AAA. Additionally, assume 100% process yields and that the average RPT for each XYZ widget on tool set AAA is 2 units. In steady state, this makes for a total RPT required per day of 10 (=5x2) units per widget of XYZ on tool set AAA. Assume we start 1 widget per day and we have 10 units of capacity, what would the cycle time be?

In a completely deterministic situation (α = 0), the cycle time is of course 1 day and the tool is 100% utilized. As the variability increases and α approaches 1, the CTM equation makes it clear that the cycle time would become very large.

Common values for α, range between 0.35 and 0.65. These can be easily be estimated by looking at the current tool utilization and cycle times.

If the business wants to run this product with a CTM of 4.0, it requires some portion of time that the tool set is available to produce, but does not have WIP. How do we incorporate that into the planning process?

We calculate a burden or uplift factor (ULF) per widget based on the target CTM and the specific characteristics of the Operating Curve for this tool set. Assume the Operating Curve for this tool set has offset=1, alpha=0.5, and m=1. Using the above equation to solve for UTIL, the required utilization to achieve the CTM target of 4.0 is 0.80 (80%). For each unit of raw capacity required, we need 1.25 units available to meet the CTM target. The value 1.25 is the uplift factor (ULF) and determined by:

ULF = 1 / tool_utilization_to_meet_cycle_time_target_from_operating_curve

If we have 250 units of capacity available per day, how many widgets can we start per day at committed cycle time? The answer is 20 = {250/ (1.25x10)}.

If the business wanted to achieve a cycle time of 3.5, how many widgets could it start per day? Using the same equation for UTIL, the utilization required to achieve a cycle time of 3.5 is 0.75 (75%). The ULF is 1.333 = 1/0.75. The maximum number of widgets per day at this cycle time commit is {250/ (1.333x10)} = 18.75. Shorter cycle time equates to reduced widget starts.

Alternatively, instead of decreasing the capacity available, we can uplift the capacity required per unit of production. For the 4.0 cycle time commit, we uplift 10 units per day per widget to 12.5 (10 x 1.25). For the 3.5 commit, the 10 units are uplifted to 13.33 (10 x 1.33). Figure 5-1 gives us the uplifted capacity required to input to the central planning process based on the cycle time the business wishes to achieve.

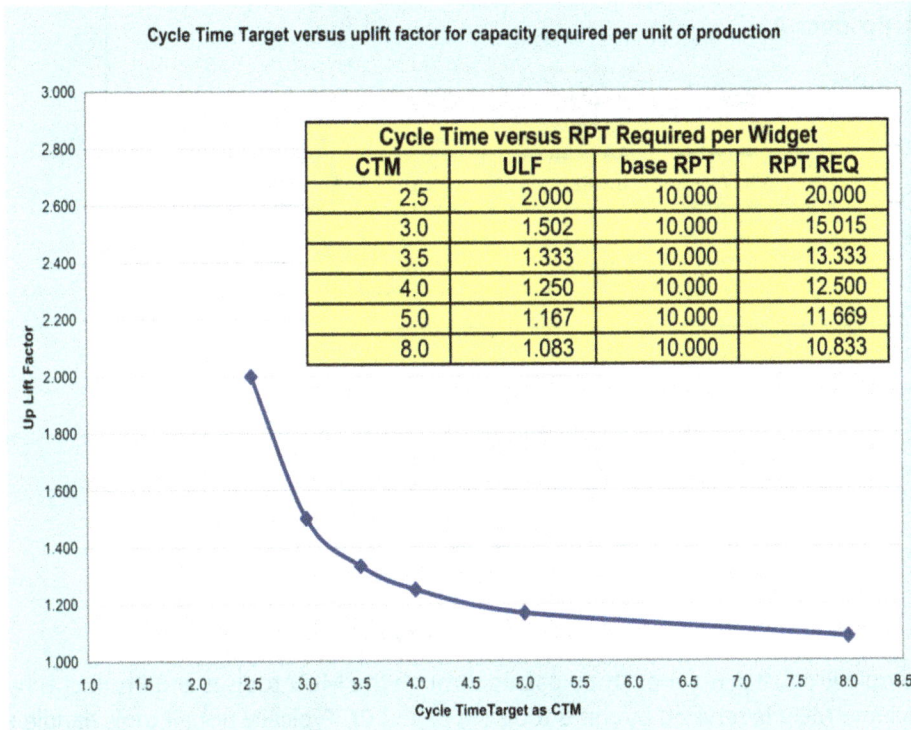

Cycle Time Target versus uplift factor for capacity required per unit of production

Cycle Time versus RPT Required per Widget			
CTM	ULF	base RPT	RPT REQ
2.5	2.000	10.000	20.000
3.0	1.502	10.000	15.015
3.5	1.333	10.000	13.333
4.0	1.250	10.000	12.500
5.0	1.167	10.000	11.669
8.0	1.083	10.000	10.833

Figure 5-1: Uplift in Capacity Required Based on Cycle Time Commit

This example illustrates that the assumption in typical central planning processes that cycle time and capacity are independent is not correct – the two are closely coupled.

Reentrant Flow

From a PSD perspective the critical insight is the repetitive use of the same core processes across the front end and back end and hence the same core equipment sets, is called "reentrant flow." This generates the complexity of partially shared operations (PSO or cascading) and deployment decisions. The following example will illustrate these issues. Figure 5-2 has a simplified set of major manufacturing activities to build one circuit layer.

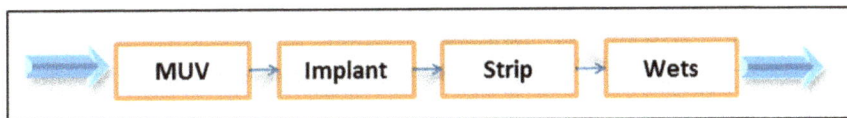

Figure 5-2 typical steps in building a circuit layer

This core set of activities is repeated multiple times to build a part and the number of iterations are different between parts. Figure 5-3 shows the number of iterations through these same steps for Product A (three passes) and Product B (two passes).

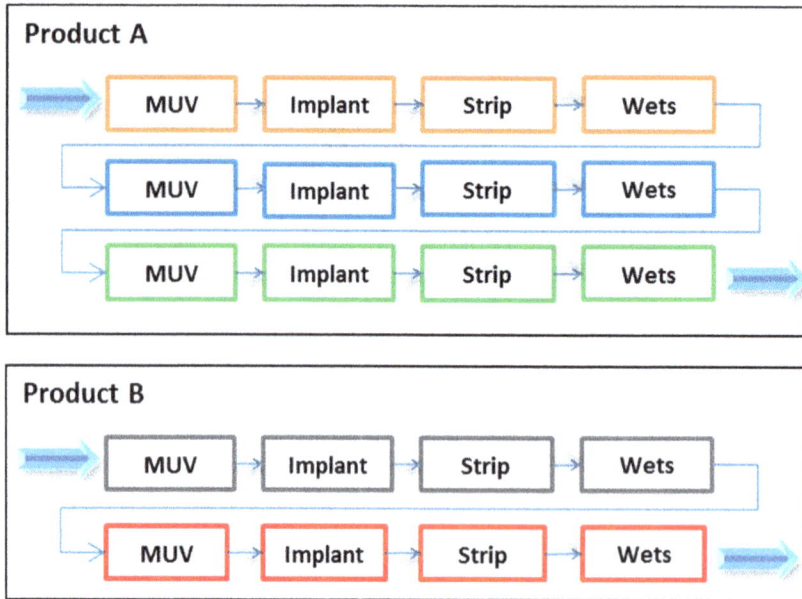

Figure 5-3: Example of reentrant flow for typical circuit construction

In this example Product A requires three passes through the MUV toolset and Product B two passes. Assume MUV is serviced by three tools (A, B, and C). Typically not all tools handle all operations (passes). Figure 5-4a demonstrates the concept of partially shared tools sets for related, but different manufacturing operations. In this example there are three tools (A, B, and C) to handle the MUV step. But not all tools can handle all operations.

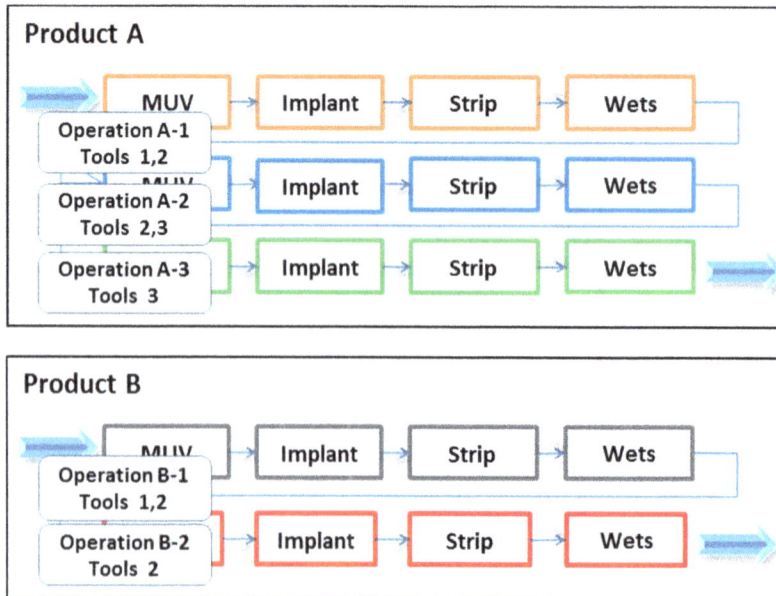

Figure 5-4a Reentrant flow with Partial Shared Operations (PSO)

In table form, the deployment decision would be represented as follows

	Tool 1	Tool 2	Tool 3	Number of tools Covering Operation
Operation A-1	1	1	0	2
Operation A-2	0	1	1	2
Operation A-3	0	0	1	1
Operation B-1	1	1	0	2
Operation B-2	0	1	0	1
Number of Operations that the tool covers	2	4	2	

Figure 5-4b: Partial Shared Operations in Table Form

The decision about which tools can handle which operations (deployment decision) is essentially determining which cells in Table 5.1 get a 1 and which get a 0. Typically a tool can service many operations, but a factory will limit the number of operations it is "allowed" to service.

Deployment has a direct impact on capacity because it limits how the workload can be assigned to the tools. However, most factories do tend to limit the tool to operation assignment because it ensure that a tool remains qualified to handle an operation by regularly servicing this operation, and it consequently reduces the work load on manufacturing engineering to maintain tool qualification

Partially Shared Operations (PSO) – Cascading

In Table 5.1 we have three tools (A, B, and C) to service seven operations (001 – 007) creating a 7 by 3 grid. A 1 indicates that this tool can service this operation and 0 it cannot. Assume there are no other tools that can service these operations and there are no other operations that these tools can service. This is called PSO or Cascade group since there are no tool/operation subsets that can be found that are independent – at some level each of these operations and tools are connected.

	Tool 1	Tool 2	Tool 3	Number of tools Covering Operation
Operation 001	1	1	0	2
Operation 002	1	1	0	2
Operation 003	0	1	1	2
Operation 004	0	0	1	1
Operation 005	0	1	1	2
Operation 006	1	0	0	1
Operation 007	1	0	0	1
Number of Operations that the tool covers	4	4	3	

Table 5.1: Partial Shared Operations in Table Form

Table 5.2 has the same tools and operations as Table 5.1, but with a different deployment allocation. In this situation there is overlap in operations between Tool 2 and Tool 3, therefore there are two PSO groups. Tool 1 and Tool 2 (and the operations they service – 1, 2, 5, 6, and 7) and Tool 3 (operations 3 and 4).

	Tool 1	Tool 2	Tool 3	Number of tools Covering Operation
Operation 001	1	1	0	2
Operation 002	1	1	0	2
Operation 003	0	0	1	1
Operation 004	0	0	1	1
Operation 005	0	1	0	1
Operation 006	1	0	0	1
Operation 007	1	0	0	1
Number of Operations that the tool covers	4	3	2	

Table 5.2: Deployment Information for 2 PSO Groups

An additional wrinkle is the raw process times for the same operation may differ between tools.

Factory Planning Basics and Challenges

	Tool 1	Tool 2	Tool 3
Operation 001	4	20	5
Operation 002	15	20	6
Operation 003	10	15	8
Operation 004	10	9	20
Operation 005	5	5	5
Operation 006	8	10	10
Operation 007	10	10	10

Table 5.3: Raw Processing time (RPT)

The critical aggregate planning question is do I have enough capacity to handle average workload? How do I mix and match the workload at each operation to the tools available to meet the most demand? Table 5.4 has the average daily workload for operation and the average capacity available for each tool.

	Tool 1	Tool 2	Tool 3	Daily Workload
Operation 001	?	?		40
Operation 002	?	?		40
Operation 003		?	?	40
Operation 004			?	40
Operation 005		?	?	50
Operation 006	?			50
Operation 007	?			50
Daily Capacity	720	1152	1296	

Table 5.4: Average Daily Demand and Capacity

The factory planner has to:

• Determine if the workload can be allocated across the tools in such a way that all of the workload can be allocate without violating capacity constraints

• If insufficient capacity exits:

  • find the optimal mix of workload that can be met without violating capacity constraints
  • find the optimal allocation that minimizes additional capacity needed and incorporates some type of fair share of pain

Table 5.5 and table 5.6 show two solutions. The first is feasible in terms of capacity but 50 units of demand are not met. The second solution meets all demand but requires more capacity than is available.

	Tool 1	Tool 2	Tool 3	Daily Workload	Demand met	Demand unmet
Operation 001	23	17		40	40	0
Operation 002	0	30		40	30	10
Operation 003		10	30	40	40	0
Operation 004		0	40	40	40	0
Operation 005		10	30	50	40	10
Operation 006	40			50	40	10
Operation 007	30			50	30	20
Daily Capacity	720	1152	1296			50

Table 5.5: Capacity Feasible Solution

	Tool 1	Tool 2	Tool 3	Daily Workload	Demand met	Demand unmet
Operation 001	23	17		40	40	0
Operation 002	0	40		40	30	0
Operation 003		10	30	40	40	0
Operation 004		0	40	40	40	0
Operation 005		10	40	50	40	0
Operation 006	50			50	40	0
Operation 007	50			50	30	0
Daily Capacity	720	1152	1296			
Required Capacity	991.8	1340	1241			
Delta	-271.8	-188	55			

Table 5.6: Demand Feasible Solution

Deployment Decision

In this example we have three tools (A, B, and C) and seven operations (operation 001 – operation 007) to handle the current WIP at this tool set. Table 5.7 shows the current deployment decisions made by the factory planner. A 1 means this tool (column) is able to service this operation (row). For now these decisions are fixed.

	Tool 1	Tool 2	Tool 3	Number of tools Covering Operation
Oper001	1	1	1	3
Oper002	1	1	0	2
Oper003	0	1	0	1
Oper004	0	0	1	1
Oper005	1	1	1	3
Oper006	1	0	0	1
Oper007	1	0	0	1
Number of Operations that the tool covers	5	4	3	

Table 5.7: Deployment Decision

Table 5.8 provides all of the key pieces of information for the model. The value in operation / tool cell is the raw process time (RPT) that the tool requires to process one lot at this operation. For example, the RPT for Tool 1 to process one lot at operation 001 is 4 time units. A missing value indicates this tool / operation combination is not currently active, corresponding to a 0 in table 5.7.

The last column (# lots) provides the number lots requiring service at that operation at this point in time. For example there are 40 lots at operation 001 waiting to get on Tool 1, 2, or 3. The last row "Capacity Available" provides the number of time units of capacity available for that tool over the service period. For example, Tool 1 has 720 units available.

	Tool 1	Tool 2	Tool 3	Number of lots in WIP waiting processing
Oper001	4	20	5	40
Oper002	15	20		30
Oper003		15		10
Oper004			20	60
Oper005	5	5	5	10
Oper006	8			200
Oper007	10			200
Capacity Available	720	1152	1296	

Table 5.8: Processing time, Available capacity, Required processing

Table 5.9 has the basic "what if" model. The value in each tool / operation cell is the business decision- the number of lots the tool is assigned to handle for this operation over some time

69

period. For example, Tool 2 will handle 30 of the lots that require service at oper001; 10 lots for oper003, and 0 lots for oper005. These 21 values (7 operations by 3 tools) represent an allocation decision.

	Tool 1	Tool 2	Tool 3	Number of lots Served	Goal	Goal Target	Delta
Oper001	0	30	10	40	=	40	
Oper002	3	0		3	=	30	-27
Oper003		10		10	=	10	
Oper004			60	60	=	60	
Oper005	0	10	0	10	=	10	
Oper006	90			90	=	200	-110
Oper007	0			0	=	200	-200
Capacity Used	765	800	1250				
Goal	≤	≤	≤				
Capacity Available	720	1152	1296				
Delta	-45	352	46				

Table 5.9: Allocation Decision and Results

The results of these decisions are found in column 5 (lots served) and row 10 (cap used). The "Number of lots Served" column is the total number of lots served for this operation across all tools. For example 40 lots at oper001 will be served – 0 on Tool 1, 30 on Tool 2, and 10 on Tool 3. The row "Capacity Used" provides how much capacity was used for each tool. This is the sum of the product of the allocation times RPT (table 16).

The last component of the model is comparing the results of the business (allocation) decision made with the goals of the factory. The factory has two goals: service as many lots as possible and use all available capacity.

This simple model enables planners to manually assess the quality of their current deployment situations and establish guidelines for manufacturing on how to allocate lots amount tools by trying different deployment decisions (Table 5.10) and or different allocation decisions (Table 5.11). Typically a planner will try different allocation decisions (leaving the deployment decision unchanged) to determine how to best to allocate WIP to tools to meet prioritized demand and then send guidelines to manufacturing. Table 5.10 shows an improved allocation (red is changed allocation decisions) plan eliminating overusing capacity on Tool 1 and reducing overall unmet demand from 337 to 330.

	Tool 1	Tool 2	Tool 3	Number of lots Served	Goal	Goal Target	Delta
Oper001	0	37	3	40	=	40	
Oper002	0	10		10	=	30	-20
Oper003		10		10	=	10	
Oper004			60	60	=	60	
Oper005	0	10	0	10	=	10	
Oper006	90			90	=	200	-110
Oper007	0			0	=	200	-200
Capacity Used	720	1140	1215				
Goal	≤	≤	≤				
Capacity Available	720	1152	1296				
Delta	0	12	81				

Table 5.10: Revised Allocation Decision

Additional complications include different lot priorities, multiple periods, and partial deployments (which occur during a phase in).

To enable the planner to model the impact of deployment decisions, we need to couple the decisions made in table 5.7 and table 5.9. They can change the deployment decision (table 5.7); then revisit the allocation decision (table 5.9); then assess the impact on the WIP waiting to; be serviced.

Note: The Planned Deployment Decision is Worthless without Execution: In this example we have made a decision to run all of the lots for oper006 on Tool 1. We have made this decision because lots at oper003 can only run on Tool 2 and lots at oper004 can only run on Tool 3. However, on the factory floor doing this type of analysis would at best be difficult. Second, from the floor's point of view, the RPT for lots at oper006 is the same for all three tools. Therefore a casual analysis would make the floor indifferent to which tool handled lots at oper006 --- with dire consequences to the factory.

6. Dispatch Challenges (assigning lots to tools)

6.1 Basic Architecture

The work by Sullivan, Kempf, Fowler, Graves, Wein, Glassey, Bitran, and many others in the 1980s and early 1990s, has explored the basics of wafer fabrication and reentrant flow. The core consideration in assigning a lot to a tool; and the basic architecture of a rule based dispatch system is now well known. The core architecture of a dispatch system and the method of obtaining and organizing knowledge have remained remarkably similar to the original work done by Sullivan [77]. Figure 6.1 provides a basic overview.

The problem has fundamentally remained true to John Fowler's 1992 (NSF Workshop October 1992 New Hampshire) description: "Dynamic Production Planning and Scheduling is needed in semiconductor manufacturing because of the complexity of the manufacturing processes including factors such as unreliable equipment, batching, reentrant flows, rework, yield loss, hot lots, combination of production, engineering and R&D lots, and varying product mix and start rates.

Future wafer FABs will probably be more automated than current factories. This will require computers to control the flow of material through the factory instead of humans.

There are, however, a few things that we have going for us.

1. Except for rework, most of the flow in a FAB is deterministic (instead of probabilistic as in a job shop)

2. The processing time per wafer or per lot or per batch is very nearly deterministic, so that once processing begins; we can get a very good prediction of when the processing will end.

3. The shop floor control systems in place in current wafer FABs provide much of the information we need in order to make good decisions

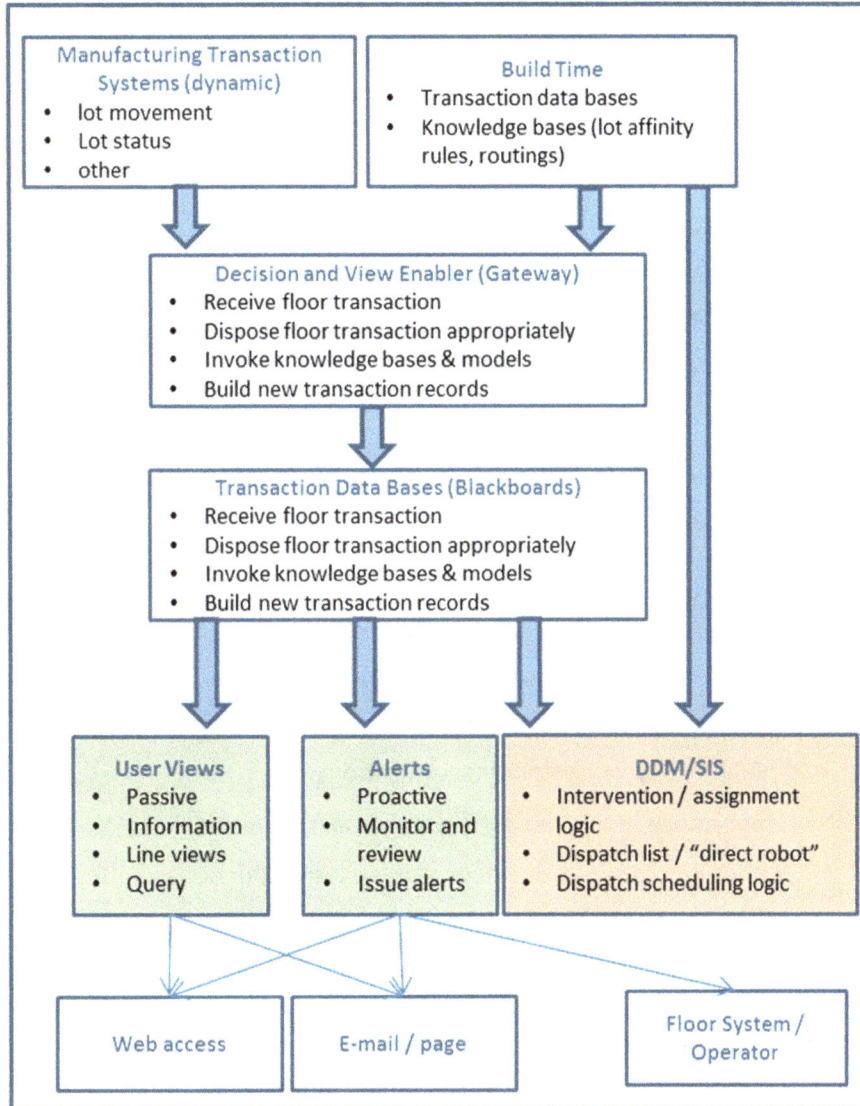

Figure 6-1: Core Structure for Real Time Dispatch Applications

6.2 Fundamentals of the Assignment Decision

The easiest way to think of dispatch is to focus on when a tool becomes free (moves from a busy state to a non-busy state or comes back on line). Lots compete to be the "next on the tool". In practice, other triggers, such as the change in status of a lot or elapsed time, can trigger the logic to review and assign lot(s) to a tool.

Figure 6-2 illustrates the basics of lots waiting for a tool at a tool set. Fundamentally this decision process must first determine which lots are eligible to run on which tools and then narrow the selection based on:

- Business guidelines (Due dates, Business Rules, dedication strategies, phase in tools);

- Process requirements (production paths, reticle availability, tool specific inhibitions); and

- Intrinsic properties of tools which significantly impact throughput (for example batch size and trains); and occasionally synchronization of events across the Fab.

We can break the key elements down as follows:

1. Tool – lot affinity

   1.1. What lots can run on this tool, what tools can handle this lot

   1.2. What are preferred tools, what are preferred lots

   1.3. Manufacturing engineering requirements

      1.3.1. Counts (avoid too many lots on certain tools)

      1.3.2. Time limits (tool requires re-qualification)

      1.3.3. Process time windows (lot must finish a sequence of steps within a time limit)

      1.3.4. Special customer specifications

2. Global importance of the lot to the supply chain or business

   2.1. Priority, customer, development versus production

3. Pacing lot movement

   3.1. Fluctuation smoothing, flow balance cycle time allocation, delta schedule, critical ratio

4. Local tool characteristics and performance

   4.1. Batching and operational trains

   4.2. Setup times

   4.3. Parallelization opportunities

   4.4. Differences in raw process time

   4.5. There may be ancillary equipment required at an operation in addition to the core toolset and labor

5. Upstream and downstream requirements

    5.1. Sending wafers to tools with limited work

    5.2. Avoiding tools with large WIP piles

    5.3. Balancing across rapidly repeating levels which use the same tool set

The core dispatch decision making activities can be divided into two primary components: guidance and judgment (figure 6-3).

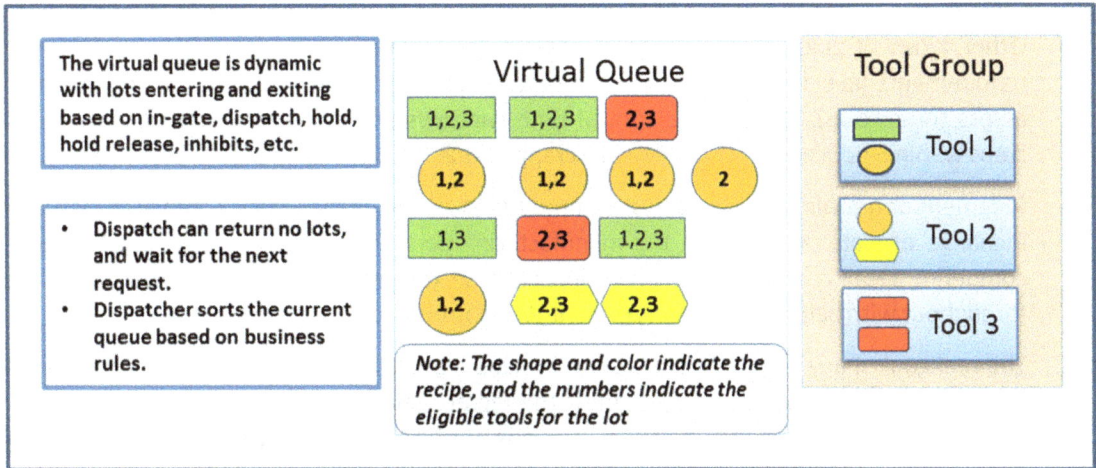

Figure 6-2: Basics of lots waiting to be dispatched

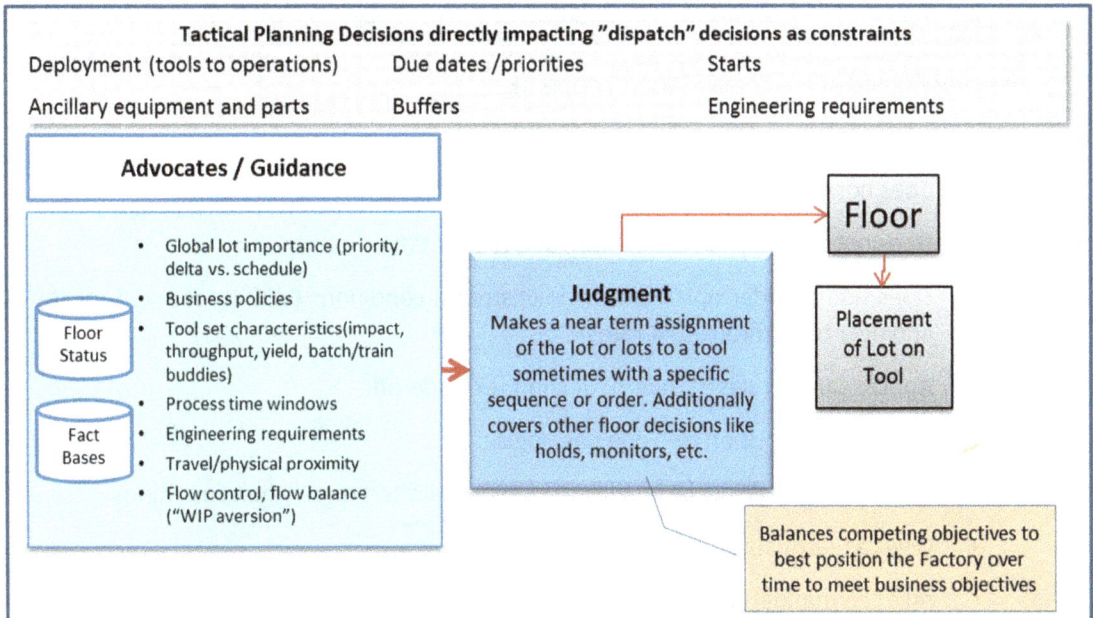

Figure 6-3: Dispatch Scheduling Framework

Guidance or advocate logic is the set of computational activities to create information that the assignment logic accesses. The calculation can be manual or automated and it normally stores the results in some sort of table structure.

The most common example is the assessment of whether a lot is ahead or behind its pace. Another example is the updating of a fact base that may contain operation – tool preference based on static information (such as difference in raw process times between tools executing the same manufacturing action) or dynamic information (the amount of time will take to set the tool up to handle this manufacturing action).

Other types of guidance include flow balance (avoid starving a tool set), manufacturing requirements (avoid running all lots of a certain type on a single tool, but distribute them across three tools), and process control time windows (lot must complete the next three steps within 5 hours or it will need to be scrapped due to contamination).

Judgment or assignment is the set of computational activities that when completed, result in a change of state or action on the manufacturing floor. The judgment logic must balance competing requirements such as meeting on time delivery, demand priorities, improving throughput with batches and trains, current WIP position, tool status, etc.

The real goal of any judgment application is make a sequence of decisions over time that in aggregate improve the future position of the factory relative to its role in the total supply chain or demand supply network. The decisions are based on impact on the future state of the factory, not based on prior events. The sum total of the prior events has resulted in the current state of the factory. All other measures are attempts to create an interim goal that can be measured and decisions made against, as a reasonable approximation of the ultimate goal. Additionally, under some circumstances, these goals can be at odds with each other or the overall good of the demand supply network.

A Simple Judgment typically:

- does not consider assignment of lots to other tools in the tool group
- Does not consider the assignment of lots over time
- Does not consider upstream or downstream conditions (WIP level and Tool status and near term throughput rates)
- Applies simple rules of thumb for complex trade-off
- written with decision tree one pass logic
- generates a single decision, through a series of filters and if-then conditions
- Gives a reasonable (though myopic) decision
- relies on manual intervention for process time windows
- Typically have to be rewritten for different WIP levels (static adjustment)

Advanced Judgment typically:

- Looks across the tool set and upstream and downstream
- Handles all process time windows
- Establishes an anticipated sequence of assignments at all tools in a tool group over time for lots at the tool or which will arrive soon
- Measures the quality of the solution and anticipates impact on factory performance
- Uses an iterative search process in judgment logic
- Dynamically adjusts for WIP levels and other business conditions
- Easily handles lots that that are intended to run at different speeds (typically given as a multiple of raw process time, referred to as CTM (cycle time multiplier) or X-Factor.

6.3 Dispatch Challenges

Simple judgment makes decisions using one-pass logic like:

- join or link
- sort and select
- scoring
- If then else
- One pass logic no looping (except accumulation), algorithm

The challenge is to move from simple judgment to advanced judgment which requires transitioning the computational tools to more advanced computational methods that can handle the advanced data structures and solution search mechanisms needed to find smarter and more responsive solutions.

6.3.1 Lots Traveling at Different Speeds

The increasing need for lots to travel at different speeds on the factory floor to meet different commitments has made simple dispatch rules such as FIFO (first in first out) and elapsed time impossible to use.

Why? Let's assume lot A03 needs travel at a speed of 3.8 (CTM) to meet its client commit date and lot A04 needs to travel at 3.3 to obtain the same result. A04 must travel faster than A03, which means it needs to absorb less wait time than A03. Inherent in the logic of "elapsed time" (lots that have been waiting the longest at a tool have the highest priority) is the wait time burden is the same for all lots, which is counter to the decision of the planners.

Assume for a moment A03 and A04 are both waiting to be processes at tool LION, A03 has been waiting at the tool for 185 hours and A04 has been waiting at the tool for 176 hours. Based on its longer elapsed time – A03 would be selected. However, if we look at the allocation of wait time burden, we see A04 at 176 is past its burden point of 175.7, while A03 at 180 is well below its burden point of 228.7.

Simple dispatch rules such as FIFO and elapsed time worked when factories made only a few products in large quantities with steady demand – a rare environment today. Factory responsiveness requires not just smarter planning, but the ability to execute which requires smarter dispatch.

6.3.2 Looking Across a Tool Set and Building a Tentative Schedule

The following example clearly illustrates the limitations of "dispatch heuristics" and the value in optimization methods to deploy dispatch scheduling. Typical "dispatch heuristics" are unable to (a) look across time, (b) look across tools at a tool set, (c) create an anticipated sequence of events at a tool set over some time horizon, (d) establish a formal metric, and (f) search for alternatives.

In top half of Figure 6-4 we have typical dispatch situation. There are three lots waiting to be processed. Two of the lots (94 and 92) are behind schedule. One of the lots is ahead of schedule (88). There are three tools in the tool group and their status is given in Figure 6-4. When tool 1 becomes available, what lot is it assigned?

	Delta vs. Schedule	Process Required		
Lot 94	behind	B	**Tool 1**	*Status*: Available to work. Currently set up on B, but can run A, B, or C
Lot 92	behind	A		
Lot 88	ahead	C	**Tool 2**	*Status*: Running. Estimated completion in 5 minutes. Currently set up on B and can run only B
• All processes are 60 minutes in length				
• All set ups take 15 minutes			**Tool 3**	*Status*: Running. Estimated completion in 45 minutes. Currently set up on A but can run A or C.

Figure 6-4: Example dispatch situation

Typically, a dispatch heuristic will only consider the tool that is available and will not perform any look-ahead to identify better alternatives. In this situation, the likely outcome of a heuristic is that it will assign lot 94 to tool A because:

- Lot 94 is behind

- The setup on tool 1 is for process B and it avoids the 15 minutes of setup time

This meets two key criteria: working on lots that are behind and improving throughput by avoiding.

As we fast forward 5 minutes, tool B becomes available. However, since it can run only process B; it is made idle because there are no lots in the queue that require process B.

This is shown graphically in figure 6-5.

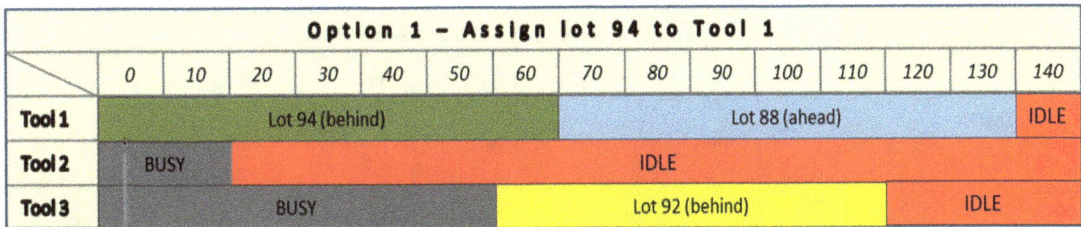

Option 1 – Assign lot 94 to Tool 1															
	0	10	20	30	40	50	60	70	80	90	100	110	120	130	140
Tool 1	Lot 94 (behind)							Lot 88 (ahead)							IDLE
Tool 2	BUSY		IDLE												
Tool 3	BUSY					Lot 92 (behind)						IDLE			

Figure 6-5: Consequences of using a simple dispatch rule

Let's return to the original decision and see how optimization can avoid the undetected collateral damage. In Figure 6-6 we see three possible solutions that are projected forward through the completion of all lots currently waiting to be serviced by this tool group. The times are we rounded up to 10 minute intervals to avoid unnecessary detail.

It is clear from the projection that option 2 and option 3 substantially outperform option 1 in terms of completion time for lots, tool utilization, or idle time

In summary:

- Option 2 generates the lowest average cycle time for all lots at the tool set
- Option 3 generates the lowest average cycle time for the "behind" lots
- Both option 2 and Option 3 complete all the work faster than option 1

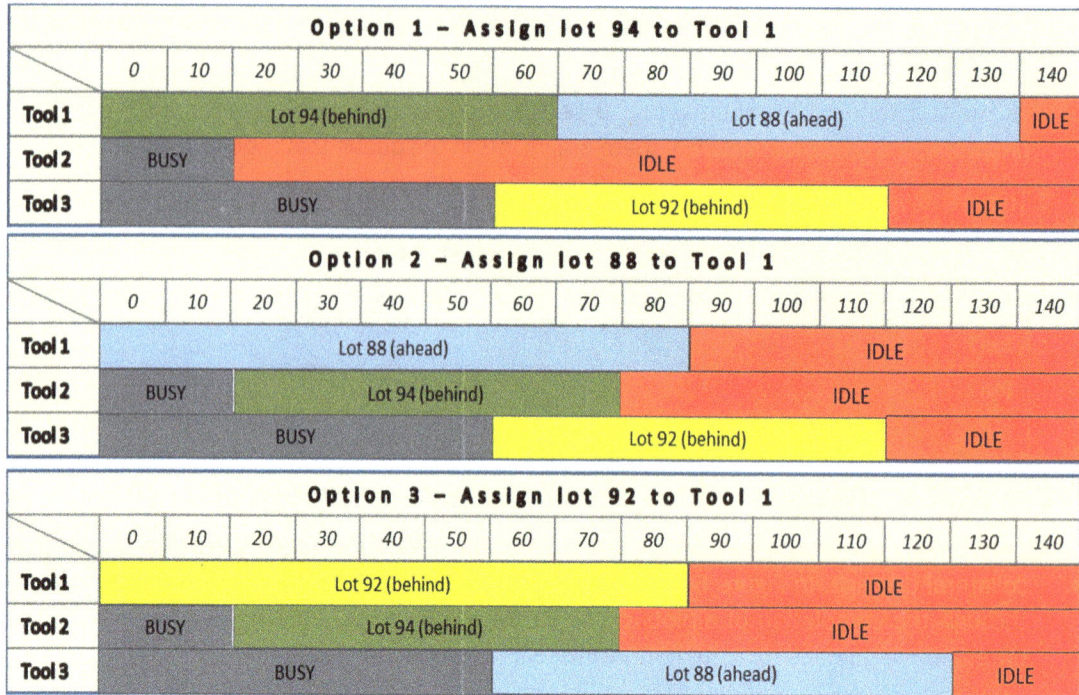

Figure 6-6: Three sequence schedule options

The assignment decision can be improved by looking at what lots are going to be available and when tools are estimated to become available. However, this additional information is only useful if it is couple with advanced computational methods to utilize the information, and clear metrics to evaluate alternatives. The best option depends on lot details (how far behind and ahead, lot priority, demand class; near term WIP conditions; and business priorities. Because these considerations are dynamic, static dispatch rules seldom achieve the capacity utilization and responsiveness of advanced computational methods.

The value of optimization goes well beyond a better assignment decisions. It provides greater visibility and enables everyone to clearly see an anticipated projected schedule for lots and tools together with anticipated tool utilization and WIP levels.

This is critical, since the assignment decision is strongly influenced by other decisions like when to qualify a tool and the movement, location, and cleaning of masks or reticles.

6.3.3 Limits of static dispatch rules

A typical goal for a FAB is to eliminate unnecessary transport of lots as part of an overall goal to minimize cycle time. A rule engine requires the rule writer to specify "how to limit" travel. It cannot simply specify a goal of limiting travel and hope the dispatch rule will achieve it.

On the other hand, an optimization procedure takes as its input the goal and then evaluates alternatives within constraints to pick the option that best meets the goals.

A typical rule would be:

"If lot is of type A, B, or C do not let it travel to another sector to be processed by a tool until it has waited at least 30 minutes in its current sector".

What the rule really wants to do is to keep a lot from traveling to another sector from its current sector if there is a tool in this sector that can handle the lot reasonably soon. There is no way to specify this general policy. So the rule specifies the "specifics" of how and hopes many times a tool will become free in the lot's current sector within 30 minutes that is a good match for this lot.

Obviously, if a fab is heavily loaded this is a higher probability event. A fab that is only moderately loaded could have a problem and the rule would need to be manually changed.

A second common limitation involves the use of rankings. Assume lots A, B, and C is behind schedule by 1.1, 1.2, and 4.8 days respectively. Lot C has the rank of 1 (most behind) since it is 3.6 days worse off than lot B. Alternatively if lot C was behind schedule by 1.3 days it would still have the rank of 1, but it just barely worse off (0.1 days) then lot B. Typically, rule based dispatch will work with the ranking and not be able make use of the richer information about the magnitude of the difference.

6.4 Process Time Windows – Avoiding Turning into a Pumpkin

Increasingly Wafer Fabrications have to "dispatch / schedule" for a process time window that impacts yield. That is once a lot is started at a certain manufacturing activity it must progress through a sequence of manufacturing activities in a certain amount of time (see Figure 6-9) to avoid a yield loss.

The typical reason is to avoid some type of contamination from exposure. These types of windows occur often in life (for example roof work, new window, plumbing that turns off the water, lane closures on highways, treating a medical condition).

For FABS they have become increasingly numerous and a nuisance over the past few years. We will call the sequence of activities in the process time window, the zone of control (ZOC).

Each lot must make it through the manufacturing steps in the ZOC within a time target. An additional complication is that each manufacturing step is handled by a different tool set. The number and type of tools in each tool set vary and generally are required to process lots that within a Zone of Control and lots that do not have a time window. There are two key decisions:

1. When to release the lot into the ZOC

2. How to assign a lot to a tool if the tool is part of a ZOC

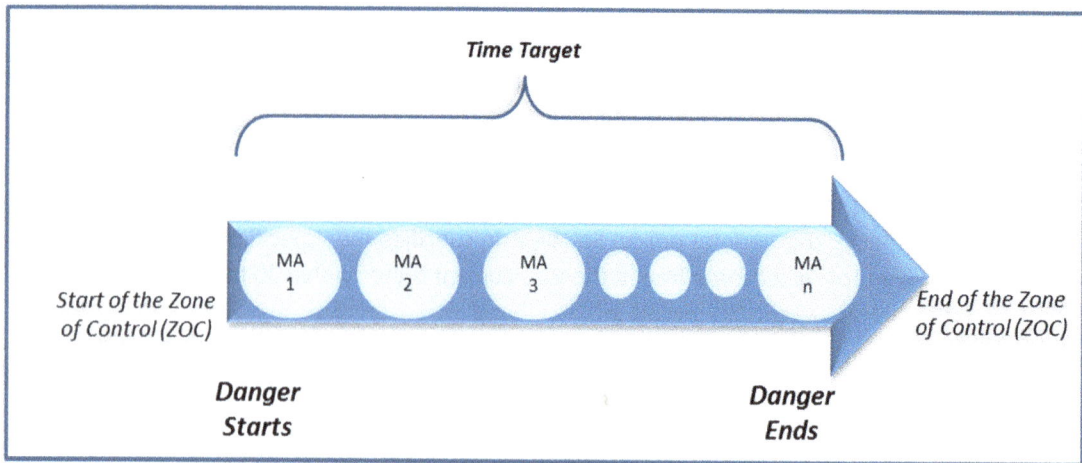

Figure 6-7 Basics of a Process Time Window Zone of Control

These decisions must consider the following objectives:

- For time window lots in the ZOC, get them onto tools in a timely manner to finish before the time limit expires

- The logic that controls the release of time windows lots into the ZOC, should determine that there is a high probability the lot will finish on time

- Avoid too many ZOC lots in the ZOC - logic that controls release must also preserve enough capacity to handle the NONZOC lots

- Balance the work load across the ZOC and NONZOC

Our observation is that the typical dispatching rules that use simplistic logic (if/then/else/sort/select) struggle to find a quality solution. They often end up being either too conservative thereby lowering capacity utilization, or run into conflicts.

Manually making these decisions is not an option. Given the number of factors involved, a typical scheduler cannot be reasonably expected to come up with good decisions consistently, without the aid of tools that can simulate decisions and present the consequences. In the absence of such tools most of the decisions are made to avoid the next perceived problem. Often, this approach simply creates other problems down the road. Of course, re-entrant flow further complicates the scheduling decisions.

7. Interaction between Central Planning, Factory Planning, and Dispatch

For the most part, planning, scheduling, and dispatch tools and processes have focused on central planning, factory planning, and dispatch separately – with at most some very loose coupling between these three areas. Dealing with these interactions is one of the key 'ongoing challenges'.

7.1 Limits of Central Planning

The primary steps in the central planning process are:

1	Create a demand statement
2	Capture the flow of materials in the demand supply network
3	Gather and collect key information from the factory 1.1. Project the completion of WIP to a decision point (often completion of the part). 1.2. Create a statement of required and available 1.3. Create a statement of lead time or cycle time to complete a new start

4	Create a model that captures key enterprise relationships (Central Planning Engine – CPE
5	Create an enterprise wide central plan by matching current and future assets with current and future demand using the CPE to create a future projected state of the enterprise and the ability to soft peg the current position of the enterprise to the projected future position. Information from the model includes: 1. a projected supply linked with exit demand 2. identification of at risk orders either to a commit date or request date 3. Synchronization signals across the enterprise 4. Capacity utilization levels

Steps 4 and 5 in the central engine planning process (CPE) are often viewed as the planning "hub" and the focus of making a firm more responsive through "smarter" engines and better (quality and timeliness) data. However, in Step 3 the factory sets the boundaries of "responsiveness". The CPE relies on the factory to provide:

1. An estimated completion date for each lot in the line (either to completion or staging point)

2. A statement of capacity available and required for each manufacturing start (typically at an aggregate level)

3. An estimated lead or cycle time to complete a start fixed for some time interval

Additionally, central planning process cannot change the due date on the lot or the lot's priority without extensive manual negotiations with manufacturing.

Central planning also has no control (and typically no knowledge) of the lot importance metric used by the factory or how it balances utilization and delivery.

Each piece of information supplied by the factory to the central planning process is "fixed" – stripped of all of the information that enables trade-offs to be made. For example:

1. Slowing one lot down to enable another lot to go faster
2. Trade-offs between cycle time and capacity available based on the operating curve
3. Redeploying tools to handle a different mix of manufacturing processes or products

Culturally, factories "dislike" change. Factories "like" steady rates of production referred to as smooth flow. This has been reinforced with various lean initiatives.

Despite the substantial forces to limit change, constant pressure from emerging market opportunities to manufacturing excursions to inaccuracies in planning (deviations between the plan and the actual) result in an ongoing sequence informal negotiations between the central planner and the factory planner to make adjustments that rely on quasi-manual decision support tools.

For example:

1. A client may need three lots four days earlier than committed and this can be accommodated simply by placing these lots on expedite.

2. The demand for product A requires 30 units of capacity of Tool Set A1 on average each day. Tool Set A1 only has 25 units of capacity available. Tool Set A2, which is not listed as a capacity option for product A, can service product A, but it runs much slower. A review of capacity utilization for Tool Set A2 indicates it will be underutilized. A decision is made to qualify Tool Set A2 to handle product A.

3. A client has had a steady order for 10 units daily of product W with a cycle time of 15 days where the key tool set is Tool Set W3. The business has been able to achieve an on time delivery rate of 97%. The client would like to increase it standing order from 10 units to 12 units. Central planning initially rejects this opportunity since the stated maximum capacity in their model for Tool Set W3 is 10. However, when the two planners look at the details of the tool set and its operating curve, the business decides it can commit to 12 per day if the cycle time is increased to 16 days (or if the OTD commit percentage is lowered).

4. Assume a client has placed an order for 5 lots of "part A" per day with a cycle time of 10 days. On average there are 50 lots of "part A" in WIP and the factory completes 5 lots per day. The due date posted on each lot is the start date plus 10 days. For example, lots started on day 6 are due on day 16. Due dates on the lots can only be changed manually by a factory planner. The factory has an abnormal set of tool outages and goes 3 days without delivering any lots – it is past due 15 lots (3x5). It has continued to start 5 lots per day. At the start of the fourth day the number of lots in the line is 65 (50 + 15 past due). On the evening of the third day the client and central planning huddle about a recovery strategy. The client determines demand has been soft for this part and agrees to "forgive" 5 lots and have the remaining 10 lots "caught" up at a pace of 1 per day (in addition to the regular 5). Therefore the new order book for this client is 6 lots per day for the next 10 days and then returns to 5 per day. Without changes to the due date on the lots in WIP, the factory continues to see 15 lots past due and will drive to "catch up" as quickly as possible. The factory may decide to delay lots for a second client to catch up all 15 past due lots in 5 days. Therefore the factory planner has to manually change the due dates on the lots to insure the factory floor has the correct guidance.

These examples make clear the opportunity for central planners to make effective use of the flexibility within the factory that is hidden. It points out the need to use the central plan as a starting point for execution rather than a fixed directive for execution.

Unfortunately, each one of these would be considered muda (wasteful) according to lean thinking and the correct response would be to eliminate them. Our experience is that you cannot "wish away" uncertainty by eliminating the processes that deal with uncertainty. You cannot eliminate snow storms by removing your snow tires.

Our contention is that in today's environment it is more effective to build tools to make to make ongoing decisions more intelligent and efficient. The goal of any firm should be to eliminate unnecessary complexity, but ignoring the complexity that remains is like tackling snow storms in the northeast with bald tires.

Additionally, such tools and processes can help to counter some of the aggregation assumptions made by the central plan. Just as the factory prefers "steady" and conservative flow, central planning often falls prey to an overly optimistic mindset that ignores some of the detailed manufacturing concerns.

Tools for improving factory/enterprise coordination fall into three groups:

1. Tools that make the central plan more visible to the enterprise. These include tools that can soft peg a lot and project its connection to the final demand, tools that project the WIP into the future, and other diagnostic tools that can simulate the effect of deviating from the plan.

2. Tactical decision models (tool deployment, allocation of cycle time or moves) that can optimize local decisions by looking at a number of alternatives simultaneously.

3. Better dispatching tools that can utilize information across the factory floor.

7.2. Committing Some Lots to run a little Faster – Collateral Impact

A common(yet manual) practice is for central planning to negotiate with factory planning to "speed" up certain lots to meet a customer request, overcome a manufacturing delay, or compensate for a planning failure. Typically, the analysis is limited and ad hoc with no systematic process. To make sure that this does not get out of hand, rules of thumb are used like:

• Number of "expedite" lots cannot exceed some fixed number N or percentage of total lots.

• The fastest an expedite lot can run is some CTM less than the normal lots. If normal lots have a CTM of 5, expedite lots might have a CTM of 3.

A closer look makes it clear the core of this decision is a reallocation of either wait-time or factory-moves that enables some lots to run faster by having others run slower over some subset of the manufacturing line and over some time duration. For example, assume the factory has five lots (LOT01 ... LOT05) in the last stage of production; each lot requires 4 moves (manufacturing actions) to complete; the maximum number of total moves (capacity) per day is 5; and the most moves a lot can have in one day is 2. If capacity is allocated "fairly," then each lot gets one move per day and each lot finishes in 4 days. Now assume the business decides LOT01 and LOT02 must finish in two days, then each needs 2 moves per days for two days, and therefore on each of these days two lots sit "idle" during these two days to enable this expedite.

Fordyce and Milne [18] develop this allocation concept in more detail focusing on wait time allocation instead of moves. In each case lots that look essentially the same are required to run at different speeds on the factory floor. The planned speedup is worthless without successful factory execution. This places a substantial burden on dispatch and precludes the use of simple methods (and lean favorites) such as FIFO (first in first out) and elapsed time. Again we see that lean and responsiveness are not always in sync.

7.3 Smarter Central Planning Through Better Modeling of Factory Capacity

As we outlined before, the central planning process requires two critical inputs from the factory: capacity (required and available) and cycle times. Since the 1980s manufacturing resource planning (MRP) and material balance equations (MBE) in optimization formulations have been the two dominant methods used in central planning [74]. In these methods the factory representation is "static" and linear. The cycle times and capacity information are fixed across some time period and handled with linear relationships.

Historically intricacies of factory tool planning (availability, deployment decisions, cascading, setup times, batching, etc.) and the dynamic interaction between equipment utilization (effective capacity) and cycle time through the operating curve have for the most part been ignored. This will not be sustainable in the future as the burden on responsiveness resulting in underutilization or delivering products late is fast becoming increasingly unacceptable.

In the previous section we described a situation where the client needed 30 units per day and the initial central planning analysis determined the maximum the factory could make was 25. The following section elaborates on the method to improve responsiveness by capturing alternative deployments of tools to manufacturing operations.

7.3.1 Revisiting Capacity Allocation – a Rabbit Out of the Hat

The core of resource allocation in central planning engines (CPE) is linking a manufacturing activity to one more resources; establishing a consumption rate for each unit of production by that manufacturing activity for the selected resource; and providing the total available capacity for the resource. In Table 7.1, we see operations 101 and 151 can be handled by Tool 1 or B. Operations 201, 202, 301, and 302 can be handled only by Tool 1. Table 7.2 tells us the available capacity for Tool 1 and Tool 2 is 1152 working minutes per time unit (for example per day).

	Resource	Consumption Rate
Operation 101	Tool 1	10
Operation 101	Tool 2	10
Operation 151	Tool 1	10
Operation 151	Tool 2	10
Operation 201	Tool 1	15
Operation 202	Tool 1	15
Operation 301	Tool 1	15
Operation 302	Tool 1	15

Table 7.1: Operation to Resource Linkage

	Consumption Rate
Tool 1	1152
Tool 2	1152

Table 7.2: Available Capacity

Assume we have a uniform start rate of one lot per day and each lot goes through each operation (101, 151, 201, 202, 301, and 302) once. In steady state each operation would need to process one lot per day. The optimal way to allocate operations to tools is to assign operations 101 and 151 to Tool 2 and the remaining four operations (201, 202, 301, and 302) assigned to Tool 1. This creates a load on Tool 2 of 20 (=10 + 10) minutes of processing and 60 (=15+15+15+15) minutes on Tool 1. Since Tool 1 has 1152 units available, the maximum number of lots per day is 19.2 (=1152/60).

If the demand rises to 30 lots, the CPE would indicate this was not feasible and mostly likely push some production out in time showing the pieces being delivered late.

However, the CPE lacks access to the tactical deployment detailed information and therefore has no method to identify better solutions. If the customer is important and his demand projected to be delivered late, the central planner will normally contact the factory planner and they will review the detailed deployment decision to "mine for capacity" by looking for opportunities to reallocate capacity to satisfy the demand on time.

The reality is that a tool can potentially handle many different operations, but at a given point in time is only actively deployed (linked) to a small subset of these operations. It is the small subset that is passed on to the CPE for its calculations.

This "reduced" deployment occurs for a number of reasons including:

1. It is physically impossible for the tool to be "actively" ready for more than a small number of operations. If we want the tool to handle an operation different than those currently selected, the tool has to be brought down for a while, "reconfigured," and brought back up.

2. There are manufacturing performance advantages to limiting the number of operations a tool is currently deployed to handle.

3. The manufacturing team often uses deployment decisions to attempt to "balance" tool load by estimating future workload.

4. The manufacturing team prefers to deploy its fastest tools to certain operations to keep total cycle time low.

5. Habit or prior practice.

The reality is that the tactical deployment decision made by manufacturing when reflected in the capacity information sent to central planning understates the flexibility of manufacturing to produce parts to meet an increase in demand. This flexibility can only be uncovered through manual intervention when central planning presses factory planning.

In our example, it might well be Tool 3 can, after being retooled, be switched from working on "gadgets" to "widgets" --- specifically it could be re-configured to handle operations 201 and 301. This is reflected in the following updated tables (7.3 and 7.4). To add to the complexity, Tool 3 may still be needed for some gadget operations.

	Resource	Consumption Rate
Operation 101	Tool 1	10
Operation 101	Tool 2	10
Operation 151	Tool 1	10
Operation 151	Tool 2	10
Operation 201	Tool 1	15
Operation 201	Tool 3	30
Operation 202	Tool 1	15
Operation 301	Tool 1	15
Operation 301	Tool 3	30
Operation 302	Tool 1	15

Table 7.3: Operation to Resource Linkage

	Consumption Rate
Tool 1	1152
Tool 2	1152
Tool 3	2000

Table 7.4: Available Capacity

Additionally, the model in the central planning engine can be enhanced to accommodate:

1. Tool – operation preferences and costs (only assign operation 301 to Tool 3 as a last resort)

2. Minimum tool usage (if operation 301 is assign to Tool 3 then it needs least 2 units of work per day to stay qualified)

3. Limitations of tool-operation pairings (Tool 3 can do work from Operation 301 or 201, but not both)

We might be tempted to state the old adage "garbage in garbage out". This is wrong and would be a failure to understand the environment that generated the "limited" but accurate capacity information. In fact, the capacity information provided by manufacturing to the central supply chain model as reflected in Tables 1 and 2 was accurate, but limited. It was limited to the current near term production requirements and the near term ability to use the tools. Tool 3 cannot be used for operations 201 and 301 until it is "retooled" and qualified. This may take a day or may take a week – but certainly manufacturing cannot afford to "retool" on a daily basis.

7.4 Supporting a Dynamic Environment

The current business environment requires that changes in the environment be incorporated into the planning process quickly. At the same time a plan is just that – a plan. Changing a plan in response to every perturbation would not do anyone any good.

Most companies can live with a plan that is created once a month. This is a natural planning cycle because most financial and accounting reporting is done with this frequency. There is also a strong argument that operational planning should be synchronized with financial operations.

For semiconductors, this is a significant issue because the time required to react is usually determined by the flexibility of the assets at the beginning of the manufacturing process. These assets are usually expensive and changing schedules quickly is either not feasible or expensive.

Recently, some companies are looking at planning in a slightly different way in an attempt to divorce updates to the plan from a rigid monthly cycle. Rather than thinking of the central planning engine (CPE) as a process, they regard the plan as an entity that evolves over the month. The demand and capacity updates are maintained in the planning data repository, as are any changes to the input data like sales and marketing input, manufacturing upsets and other external deviations from the plan. These inputs are not entered on any cycle but are updated when they occur.

Three types of changes are monitored. The planning database is constantly updated with changes from sales, customers, and customer service representatives. It is also updated with the manufacturing performance to the plan, and the current factory capacity.

These changes and deviations from the monthly plan are continuously monitored, with significant changes automatically distributed through dashboards and warning messages.

The key feature of this approach is that the central planning engine is executed when accumulated changes exceed a particular threshold, rather than on a rigid monthly cycle.

Of course, each change needs to be evaluated for the potential impact on the plan. Some of this can be automated with simple rules like "ignore all demand changes to non-critical products", or "ignore all changes to supply less than 5%", and so on. In our experience, more than 70 to 80% of all changes can be handled in an automated manner.

Significant changes that exceed predefined thresholds require a separate review process for changes.

Let's take the case of a customer that seeks chapter 11 bankruptcy protections. If this is not a significant customer (as defined by a threshold value for revenue) then a routine warning message is communicated. If this is a significant customer, then it might affect the supply plan for the products that the customer buys and a review process to evaluate the significance of this event is initiated.

The goal of the review process is to determine if the event can be handled by changes within the current plan, or if a new plan has to be created. This review process cannot be automated. It is usually assisted by a system that can simulate changes against the current plan, but it cannot be totally automated because the changes required depend on when in the month the event occurs, the business climate, and other considerations.

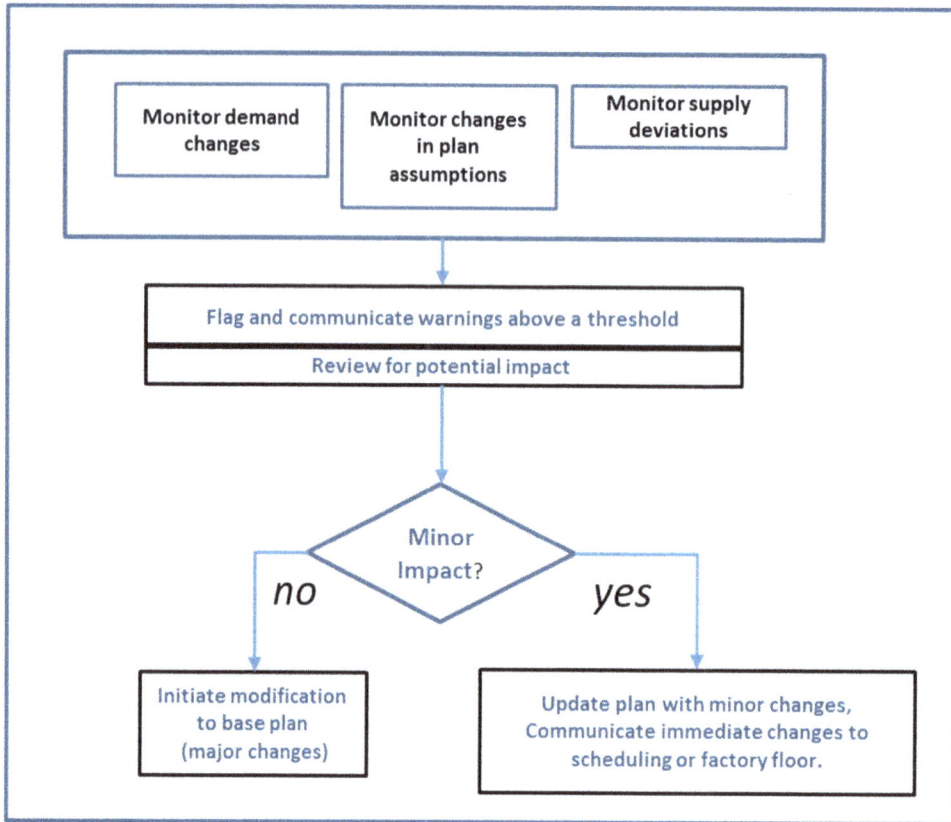

```
┌──────────────────────────────────────────────────────────────┐
│  ┌────────────────────────────────────────────────────────┐  │
│  │ ┌──────────────┐ ┌──────────────┐   ┌──────────────┐    │  │
│  │ │Monitor demand│ │Monitor changes│  │Monitor supply│    │  │
│  │ │   changes    │ │   in plan     │  │  deviations  │    │  │
│  │ │              │ │ assumptions   │  │              │    │  │
│  │ └──────────────┘ └──────────────┘   └──────────────┘    │  │
│  └────────────────────────────────────────────────────────┘  │
│         ┌──────────────────────────────────────────┐          │
│         │ Flag and communicate warnings above a threshold │    │
│         ├──────────────────────────────────────────┤          │
│         │       Review for potential impact         │          │
│         └──────────────────────────────────────────┘          │
│                       ◇ Minor Impact? ◇                        │
│              no                         yes                    │
│  ┌──────────────────┐        ┌──────────────────────────────┐ │
│  │Initiate modification│      │Update plan with minor changes,│ │
│  │  to base plan       │      │Communicate immediate changes to│
│  │ (major changes)     │      │ scheduling or factory floor.  │ │
│  └──────────────────┘        └──────────────────────────────┘ │
└──────────────────────────────────────────────────────────────┘
```

Such a simulation system looks at the current status (supply, demand, transportation, and inventory) and evaluates the consequence of different actions. In our case, we will want to know the financial cost of shipping the remaining material assuming that we will not get paid, the opportunity costs if we can use the capacity committed to this customer to make other products, and so on. For example, if the product for this customer has already been manufactured, and the product is so specialized that there is no other significant market for it, the right decision might be to ship the product anyway; but plan on no additional production for this product when the monthly plan is revised next month.

In our example, if the customer has taken most of his forecast for the month, then the current plan would likely not be changed but the information would be used in the next planning cycle. If on the other hand, significant shipments are scheduled, then the situation might be handled through a direct intervention at the tactical scheduling level to delay the shipments until additional credit checks are complete.

On the other hand, if significant manufacturing resources are yet to be allocated to meet this customer's demand, then it might indeed be reasonable to revise the current monthly plan. Where and how to intervene cannot be automated, but the process to deal with changes can and should be.

The response depends on many factors that might not all be captured in systems, and for this reason can be totally automated by a computer system.

Bibliography

1. Bermon, S. and Hood, S. (1999), "Capacity Optimization Planning System (CAPS)," Interfaces, Vol. 29, No. 5, pp. 31 – 50.

2. Bitran, G. and Tirupati, D. (1989), "Tradeoff Curves, Targeting and Balancing in Manufacturing Networks," Operations Research, Vol. 37, No. 4, pp. 547- .555

3. Bixby, R,, R. Burda, and D. Miller. 2006. Short-interval detailed production scheduling in 300mm semiconductor manufacturing using mixed integer and constraint programming" in Semiconductor fabtech – 32nd edition, <WWW.FABTECH.ORG,> 34-40

4. Bixby, R. 2009, "From Planning to Operations: The Ever-Shrinking Optimization Time Horizon", Workshop on Combinatorial Optimization at Work, Conference, Sep 21 - Oct 9, 2009 at ZIB, Berlin, Germany, co-at-work.zib.de/berlin2009/downloads/2009-09-25/2009-09-25-1600-BB-Real_Time_Applications.pdf

5. Chen, H., Harrison, M., Mandelbaum, A., Ackere, A., and Wein, L. (1988), "Empirical Evaluation of a Queuing Network Model for Semiconductor Wafer Fabrication," Operations Research, Vol. 36, No. 2, pp. 202-215.

6. Connors, D., Feigin, G., and Yao, D., 1996, "A queuing network model for semiconductor manufacturing", IEEE Transactions on Semiconductor Manufacturing, 9 (3), 412-427

Bibliography

7.   Consilium, 1988. Rule Based Dispatch User's Guide, Mtn View, CA

8.   Dabbas, R. and J. Fowler 2003. A New Scheduling Approach Using Combined Dispatching Criteria in Wafer Fabs, IEEE Transactions on Semiconductor Manufacturing, 16:3 501-510

9.   Dangat, G.S., Milne, R.J., and Orzell, R.A. (2000), "Method to Provide Common Support for Multiple Types of Solvers for Matching Assets with Demand in Microelectronics Manufacturing," U.S. Patent 6,041,267 (3/21/2000).

10.   Davenport, T. (2006), "Competing on Analytics," Harvard Business Review, January 2006, pp. 1 – 9.

11.   Dennis, P. (2007), Lean Production Simplified, Productivity Press, NY

12.   Denton, B., Hedge, S., and Orzell, R.A. (2004), "Method of Calculating Low Level Codes for Considering Capacities," U.S. Patent 6,584,370 (5/18/2004).

13.   Denton, B., Forrest, J., and Milne, R.J. (2005), "A Method for Considering Hierarchical Preemptive Demand Priorities in a Supply Chain Optimization Model," U.S. Patent Application: 2005-0171828, also, IBM docket: BUR9-2003-0198US1.

14.   Denton, B. and Milne, R.J. (2006a), "Method for Optimizing Material Substitutions within a Supply Chain," U.S. Patent 6,983,190 (1/3/2006)

15.   Denton, B., Forrest, J., and Milne, R.J. (2006), "Methods for Solving a Mixed Integer Program for Semiconductor Supply Chain Optimization at IBM," Interfaces, Vol. 36, No. 5, September – October 2006, pp. 386 – 399

16.   Fargher, H. and R. Smith, 1994. Planning in a Flexible Semiconductor Manufacturing Environment, Chapter 19 in Intelligent Scheduling, edited by Mark Fox and Monte Zweben, Morgan Kaufman Publishers

17.   Fordyce, K, Bixby, R, and Burda, R. (2008), "Technology That Upsets The Social Order – A Paradigm Shift In Assigning Lots To Tools In A Wafer

Bibliography

Fabricator – The Transition From Rules To Optimization," Proceedings of the 2008 Winter Simulation Conference

18. Fordyce, K., Wang, C-T, Chang, C., Degbotse, A., Denton, B., Lyon, P., Milne, R.J., Orzell, R., Rice, R., and Waite, J. (2011a), "The Ongoing Challenge—Creating an Enterprise-wide Detailed Supply Chain Plan for Semiconductor and Package Operations," chapter 14 in Planning Production and Inventories in the Extended Enterprise: A State of the Art Handbook, Volume 2, edited by Kempf, Keskinocak, and Uzsoy

19. Fordyce, K., Milne, R.J, et al (2011b), "Improving Factory Responsiveness—an Optimization Model for Wait Time Allocation," working paper, fordyce@us.ibm.com, jmilne@clarkson.edu

20. Fordyce, K., Milne, R.J, et al (2011c), "Basics of the Operating Curve – Classical Planning Meets Its Uncertainty Principle," working paper, fordyce@us.ibm.com, jmilne@clarkson.edu

21. Fordyce, K. Milne, R.J, et al (2011d), "Dynamically Linking Cycle Time Decision with Capacity Required Inputs in Central Planning Models," working paper, fordyce@us.ibm.com, jmilne@clarkson.edu

22. Fordyce, K. Milne, R.J, et al (2011e), "Using Optimization Methods to Help with Deployment Decisions in Factory Planning," working paper, fordyce@us.ibm.com, jmilne@clarkson.edu

23. Fowler, J. W., Brown, S., Gold, H., and Schoemig, A., 1997, "Measurable improvements"

24. in cycle-time-constrained capacity", Proceedings of the 6th IEEE/UCS/SEMI International Symposium on Semiconductor Manufacturing (ISSM), San Francisco, CA

25. Fowler, J. W. and Rose, O., 2004, "Grand challenges in modeling and simulation of complex manufacturing systems," Simulation, vol. 80, no. 9, pp. 469–476, 2004

26. Fox, B. and Kempf, K (1985), Complexity, Uncertainty, and Opportunistic Scheduling, Proceedings of the IEEE Second Conference on Artificial

Intelligence Applications: The Engineering Knowledge Based Systems, Miami, Florida, 487-492.

27.   Fox, M. 1987. Constraint-Directed Search: A Case Study of Job-Shop Scheduling, Morgan Kaufman Publishers, Los Altos, Ca.

28.   Galbraith, J. (1973), Designing Complex Organizations, Addison-Wesley.

29.   Glassey, C. et al, 1996. Control Rules for Production Control Semiconductor Fabs, .IEEE Transactions on Semiconductor Manufacturing, 9:4 536-549.

30.   Glassey, C. and W. Weng 1991. Dynamic Batching Heuristic for Simultaneous Processing, IEEE Transactions on Semiconductor Manufacturing, 4:2 77-82.

31.   Goldman, S. (2004), "Science in the Twentieth Century," Great Courses on CD by the Teaching Company, Chantilly, VA.

32.   Glover, F., Jones, G., Karney, D., Klingman, D., and Mote, J. (1979), "An Integrated Production, Distribution, and Inventory Planning System," Interfaces, Vol. 9, No. 5, pp. 21 – 35.

33.   Govind, N. et al 2007, "Integrated Targeting, Near Real-Time Scheduling, and Dispatching with Automated Execution in Semiconductor Manufacturing", 2007 IEEE/SEMI Advanced Semiconductor Manufacturing Conference, pp. 87-92.

34.   Graves, S., H. Meal, D. Stefek, and A. Zeghmi, 1983. .Scheduling of Re-entrant Flow Shops, Journal of Operations Management, 3: 197-203

35.   Graves, R.J., Konopka, J.M., and Milne, R.J. (1995), "Literature Review of Material Flow Control Mechanisms," Production Planning and Control, Vol. 6, No. 5, 395 – 403.

36.   Gurnani, H., R. Anupindi, and R. Akella 1992. Control of Batch Processing Systems in Semiconductor Wafer Fabrication Facilities, IEEE Transactions on Semiconductor Manufacturing, 5:4. 319-327

Bibliography

37. Hackman, S.T. and Leachman, R.C. (1989), "A General Framework for Modeling Production," Management Science, Vol. 35, No. 4, pp. 478 – 495.

38. Hegde, S.R., Milne, R.J., Orzell, R.A., Pati, M.C., and Patil, S.P. (2004), "Decomposition System and Method for Solving a Large-Scale Semiconductor Production Planning Problem," United States Patent No. 6,701,201 B2.

39. Hopp, W. and Spearman, M. (2008), Factory Physics, 3rd Edition, McGraw-Hill, NY

40. Horn, G. and Podgorski, W. (1998), "A focus on cycle time-vs.-tool utilization "paradox" with material," Advanced Semiconductor Manufacturing Conference and Workshop Proceedings, pp. 405 – 412

41. IBM white paper G299-0906-00 (2006), "Collaboration with IBM E&TS Helps ADI Stay ahead of Customer Demand," IBM Corporation, 1133 Westchester Avenue, White Plains, NY 10604, U.S.A.

42. Ignizio, J. 2009, Optimizing Factory Performance, McGraw-Hill, NY

43. Johri, P. 1989, Dispatching in an Integrated Circuit Wafer Fabrication Line, Proceedings, Winter Simulation Conference, 918-921.

44. Johri, P. 1993, "Practical issues in scheduling and dispatch", Journal of Manufacturing

45. Systems, 12 (6), 474-485

46. Karmarkara, U., 1993 "Manufacturing lead times, order release and capacity loading", chapter 6 in Handbooks in Operations Research and Management Science Vol. 4 – Logistics of Production and Inventory, edited Henderson and Nelson, Pages 287-329

47. Kempf, K. 1989, Manufacturing Scheduling: Intelligently Combining Existing Methods, in Working Notes of AAAI AI in Manufacturing Symposium, M. Fox editor, AAAI, 445 Burgess Drive Menlo Park, CA 94025-3496.

Bibliography

48.   Kempf, K. 1989., Manufacturing Planning and Scheduling: Where We Are and Where We Need to Be, Proceedings of the Fifth IEEE Conference on AI Applications, IEEE Computer Society Press, Los Alamitos, CA. 13-19.

49.   Kempf, K, Pape,D., Smith, S and Fox, B (1991). Issues in the Design of AI Based Schedulers, AI Magazine, 11:5, 37-45.

50.   Kempf, K (1994), "Intelligent Scheduling Semiconductor Wafer Fabrication," Chapter 18 in Intelligent Scheduling, edited by Mark Fox and Monte Zweben, Morgan Kaufman Publishers, 473-516.

51.   Kempf, K. (2004), "Control-Oriented Approaches to Supply Chain Management in Semiconductor Manufacturing," Proceedings of the 2004 American Control Conference, Boston, MA, pp. 4563 – 4576.

52.   Kempf, K. 2007, "Complexity and the Enterprise – The Illusion of Control", Intel Corp,

53.   Leachman, R., Benson, R., Liu, C., and Raar, D. (1996), "IMPReSS: An Automated Production Planning and Delivery-Quotation System at Harris Corporation - Semiconductor Sector," Interfaces, Vol. 26, No. 1, pp. 6 – 37.

54.   Lee C., R. Uzsoy, L. Martin-Vega and P. Leonard, 1991. Production Scheduling Algorithms for a Semiconductor Test Facility, IEEE Transactions on Semiconductor Manufacturing, 4: 271-280.

55.   Lee C., R. Uzsoy, and L. Martin-Vega 1992, Efficient Algorithms for Scheduling Semiconductor Burn-In Operations, Operations Research, 40:4, 764-775.

56.   Lin, G.1992. "An Opportunistic Price-Based Multiple Resource and Part Scheduling," Ph.D. dissertation, School of Industrial Engineering, Purdue University,

57.   Lin, G., Ettl, M., Buckley, S., Yao, D., Naccarato, B., Allan, R., Kim, K., Koenig, L. (2000), "Extended Enterprise Supply Chain Management at IBM Personal Systems Group and Other Divisions," Interfaces, 30(1), pp. 7 – 25.

Bibliography

58.   Little, J. (1970), "Models and Managers: The Concept of a Decision Calculus," Management Science, Vol. 16, No. 8, pp. B-466 – B-485.

59.   Little, J. (1992), "Tautologies, Models and Theories: Can We Find "Laws" of Manufacturing?," IIE Transactions, Vol. 24, No. 3, pp. 7 – 13.

60.   Liu, J., Yang, F., Wan, H., and Fowler, J. (2010), "Capacity Planning Through Queuing Analysis and Simulation-based Statistical Methods: a Case Study for Semiconductor Wafer FABS," web.ics.purdue.edu/~hwan/docs/IJPR

61.   Lyon, P., Milne, R.J., Orzell, R., Rice, R. (2001), "Matching Assets with Demand in Supply-Chain Management at IBM Microelectronics," Interfaces, 31:1, pp. 108 – 124.

62.   Orzell, R., Patil, S., and Wang, C. (2004) "Method for Identifying Product Assets in a Supply Chain Used to Satisfy Multiple Customer Demands," U.S. Patent 20050177465A1 (10/17/2004).

63.   Martin-Vega. L. et al, 1989, "Apply Just in Time at a Wafer Fab case study", IEEE Transactions on Semiconductor, Vol. 2, No. 1,16-22.

64.   Martin, D. (1999), "Capacity and cycle time-throughput understanding system (CAC-TUS)," Advanced Semiconductor Manufacturing Conference and Workshop Proceedings, pp. 127 - 131

65.   Monch, L., Fowler, J., Stéphane Dauzère-Pérès, S., Mason S., and Rose, O. 2011, "A survey of problems, solution techniques, and future challenges in scheduling semiconductor manufacturing operations", Journal of Scheduling, DOI: 10.1007/s10951-010-0222-9

66.   Morrison, J., Campbell, B., Dews, E. and LaFreniere J. 2005, "Implementation of a fluctuation smoothing production control policy in IBM's 200mm wafer FAB", Proceedings of the joint 44th IEEE Conference on Decision and Control and the European Control Conference, Seville, Spain, 2005.

67.   Morrison, J., Dews, E. and LaFreniere J. 2006, "Fluctuation smoothing production control at IBM's 200mm wafer fabricator: Extensions, application and the multi-flow production index (MFPx)", Proceedings of

Bibliography

the 2006 IEEE/SEMI Advanced Semiconductor Manufacturing Conference, Boston, MA, May 2006.

68.   Morrison, J. and Martin, D. (2006), "Cycle Time Approximations for the G/G/m Queue Subject to Server Failures and Cycle Time Offsets with Applications," ASMC 2006 Proceedings, pp. 322

69.   Orlicky, J. (1975), Material Requirements Planning: The New Way of Life in Production and Inventory Management, McGraw-Hill, New York.

70.   Pogge, R. (2009), "Real World Relativity", www.astronomy.ohiostate.edu/~pogge/Ast162/Unit5/gps.html

71.   Pfund, M., Mason, S., and Fowler, J. (2006). "Dispatching and scheduling in semiconductor manufacturing: state of the art and survey of needs". in J. Herrmann (Ed.), Handbook of production scheduling (pp. 213–241). Springer-Verlag

72.   Robinson, J. K., Fowler, J. W. and Bard, J. F., 1995, "The use of upstream and downstream information in scheduling semiconductor batch operations", International Journal of Production Research, 33 (7), 1849-1870

73.   Savell, D,, R. Perez, and S. Koh, 1989. Scheduling semiconductor wafer production: an expert system implementation", IEEE Expert 4:3 9-15

74.   Shirodkar,S. and Kempf, K (2006), "Supply Chain Collaboration Through Shared Capacity Models", Interfaces, Vol. 36, No. 5, pp. 420-432

75.   Shobrys, D. (2003), "History of APS," Supply Chain Consultants (www.supplychain.com), 460 Fairmont Drive, Wilmington, DE 19808, U.S.A.

76.   Simon, H.A. (1957), Administrative Behavior, Second Edition, The Free Press, New York.

77.   Singh, H. (2009a), Supply Chain Planning in the Process Industry, Supply Chain Consultants, Wilmington, De.

78.   Singh, H. (2009b), Practical Guide for Improving Sales and Operations Planning , Supply Chain Consultants, Wilmington, De.

79. Singh, H. (2010), "Effectively Balancing Demand and Supply Forecasting in a Rebounding Economy," Supply Chain Consultants, Wilmington, DE.

80. Sullivan, G. (1990), "IBM Burlington's Logistics Management System (LMS)," Interfaces, Vol. 20, No. 1, pp. 43 – 61.

81. Sullivan, G. (1994), "Logistics Management System (LMS): Integrating Decision Technologies for Dispatch Scheduling in Semiconductor Manufacturing," Intelligent Scheduling, Morgan Kaufman Publishers, pp. 473 – 516.

82. Sullivan G. (1995), "A Dynamically Generated Rapid Response Fast Capacity Planning Model for Semiconductor Fabrication Facilities," The Impact of Emerging Technologies on Computer Science and Operations Research, Kluwer Academic Publishers, Boston (presented at Winter 1994 Computers and Operations Research Conference).

83. Tayur, S., Ganeshan, R., and Magazine, M. (1998), Quantitative Models for Supply Chain Management, Kluwer Academic Publishers, Boston, MA.

84. Uzsoy, R., Lee, C., and Martin-Vega, L.A. (1992), "A Review of Production Planning and Scheduling Modules in the Semiconductor Industry, Part 1: System Characteristics, Performance Evaluation, and Production Planning," IIE Transactions, Scheduling Logistics, 24(4), pp. 47 – 60.

85. Uzsoy, R., Lee, C., and Martin-Vega, L.A. (1994), "A Review of Production Planning and Scheduling Modules in the Semiconductor Industry, Part 2: Shop Floor Control," IIE Transactions, Scheduling Logistics, 26(5), pp. 44 – 55.

86. Wein, L. 1988, "Scheduling Semiconductor Wafer Fabrication", IEEE Transactions on Semiconductor Manufacturing, Vol. 1, No. 3., 115-130.

87. Wolfson, R. (2000), "Einstein's Relativity and the Quantum Revolution," Great Courses on CD by the Teaching Company, Chantilly, VA.

88. Woolsey, G. (1979), "Ten Ways to Go Down with Your MRP," Interfaces, Vol. 9, No. 5, pp. 77 – 80.

Bibliography

89. Zisgen, H. (2005), "EPOS – Stochastic Capacity Planning for Wafer Fabrication with Continuous Fluid Models," IBM Global Engineering Services, Decision Technology Group Mainz, Germany.

90. Zisgen, H., Ments, I., Wheeler, B, and Hanschle, T., (2008), "A Queuing Network Based System to Model Capacity and Cycle Time For Semiconductor Fabrication," Proceedings of the 2008 Winter Simulation Conference.

91. Zisgen, H., Brown, S, Hanschke, T., Meents, I., and Wheeler, B (2010) "Queuing Model Improves IBM's Semiconductor Capacity and Lead-Time Management," Interfaces, Vol. 40, No. 5, pp. 397-407

92. Miller, D. 1990, "Simulation of a Semiconductor Manufacturing Line", Communications of the ACM, Vol. 33, No. 10, pp. 98-108.

93. Burda, R., Degbotse, A., Denton, B., Fordyce, K., and Milne, R.J, "Method, system, and computer program product for controlling the flow of material in a manufacturing facility using an extended zone of control," U.S. patent 7,305,276.

94. Burda, R., Degbotse, A., Denton, B., Fordyce, K., and Milne, R.J., "Method of Release and Product Flow Management for Process Time Windows in a Manufacturing Facility", U.S. Patent 7477958.

www.ingramcontent.com/pod-product-compliance
Lightning Source LLC
Chambersburg PA
CBHW082106210326
41599CB00033B/6607